THE MAGELLANIC CLOUDS

ASTROPHYSICS AND SPACE SCIENCE LIBRARY

A SERIES OF BOOKS ON THE RECENT DEVELOPMENTS

OF SPACE SCIENCE AND OF GENERAL GEOPHYSICS AND ASTROPHYSICS

PUBLISHED IN CONNECTION WITH THE JOURNAL

SPACE SCIENCE REVIEWS

VOLUME 23

THE MAGELLANIC CLOUDS

A EUROPEAN SOUTHERN OBSERVATORY PRESENTATION:
PRINCIPAL PROSPECTS, CURRENT OBSERVATIONAL
AND THEORETICAL APPROACHES,
AND PROSPECTS FOR FUTURE RESEARCH

BASED ON THE SYMPOSIUM ON THE MAGELLANIC CLOUDS
HELD IN SANTIAGO DE CHILE, MARCH 1969,
ON THE OCCASION OF THE DEDICATION
OF THE EUROPEAN SOUTHERN OBSERVATORY

Edited by

ANDRÉ B. MULLER

European Southern Observatory, Hamburg, W. Germany

SPRINGER-SCIENCE+BUSINESS MEDIA, B.V.

Library of Congress Catalog Card Number 73-154743

ISBN 978-94-015-6908-8 ISBN 978-94-015-6906-4 (eBook)
DOI 10.1007/978-94-015-6906-4

Originally published by D. Reidel Publishing Company, Dordrecht, Holland in 1971
Softcover reprint of the hardcover 1st edition 1971

PREFACE

On March 28 and 29, 1969, at the occasion of the dedication of the European Southern Observatory, some 90 astronomers from all over the world gathered at the ESO headquarters at Santiago de Chile for discussing problems of the Magellanic Clouds. They came from Argentina, Australia, Chile, Mexico, South Africa and the United States as well as from Europe; these latter, naturally, mostly from the member states of ESO.

The choice of the subject was an obvious one. When erecting the European Southern Observatory as a joint effort in European astronomy, it was agreed from the beginning that the field of research should be the southern sky, so far hardly explored with large telescopes. Among the objects to be investigated, the Magellanic Clouds rank highest, together with the galactic centre region and the southern spiral structure. Being located ten times closer than the nearest large stellar systems accessible to northern observers, and containing a stellar population ranging in age from the oldest down to the stage of star formation, the Clouds provide an ideal laboratory for research on current problems in astrophysics.

Yet, most of the northern observational astronomers were hardly acquainted with the Magellanic Clouds; naturally, they are used to think in terms of research projects that can be conducted at their observatories. A survey of the status of knowledge and research on the Clouds therefore appeared in order now that the first – medium sized – telescopes of ESO came into operation.

Accordingly, a major part of the symposium report consists of review papers. They are followed by contributions on the current observational and theoretical approaches; and some papers indicating the prospects for future work conclude the volume. For reasons of presentation, the order in which the papers appear here differs somewhat from the one in which they were given at the symposium.

In sponsoring the symposium and publishing the present report, the European Southern Observatory hopes to have contributed to a new impetus in research on the Clouds, at a time when the erecting of new large southern telescopes or their design opens up an area of intensive investigations of these fascinating objects. It is a pleasure to thank here again all contributors to the symposium for their share in this project, and especially Dr A. B. Muller for his editorial work.

Hamburg, April 1971

A. BLAAUW
General director, ESO

EDITOR'S NOTE

In order to present an up to date contribution to the study of the Magellanic Clouds, authors who reported during the ESO symposium in March 1969 on then recent observations have matched their contributions with the present state of investigations.

A few contributions to the symposium are not published in this work: The paper of Fehrenbach: 'La détection et la mesure des vitesse radiales pour la reconnaissance des étoiles des Nuages de Magellan', reporting on the photographic material, collected with the Radial Velocity Astrograph. This paper will be published in one of the forthcoming ESO Bulletins. The contribution of S. Gaposhkin 'Seventy-Six Eclipsing Variables on and in the Large Magellanic Cloud and its Topography' containing extensive tables with observational data, will be submitted for publication in the Supplement Series of *Astronomy and Astrophysics*. The paper of E. Maurice 'A. Description du spectrographe Cassegrain "Chilicass" utilisé à La Silla – B. Etude de deux étoiles du Grand Nuage de Magellan présentant des raies d'émission du fer' has been published in the ESO Bulletin No. 7 as far as part A is concerned and part B has been published in *Astron. Astrophys.* **3**, 323, 1969. Rudolph's contribution on photometric scanning of the LMC being a description of his first attempt of cloud scanning. The paper of Demarque 'The Evolution of Low Mass Stars' could not be presented here for special reasons. Blanco reported on the development of the AURA Observatory on Tololo. His contribution will be published in one of the forthcoming ESO Bulletins.

I wish to thank all authors for their kind co-operation which made the editorial work a pleasant task.

Hamburg, April 1971 ANDRÉ B. MULLER

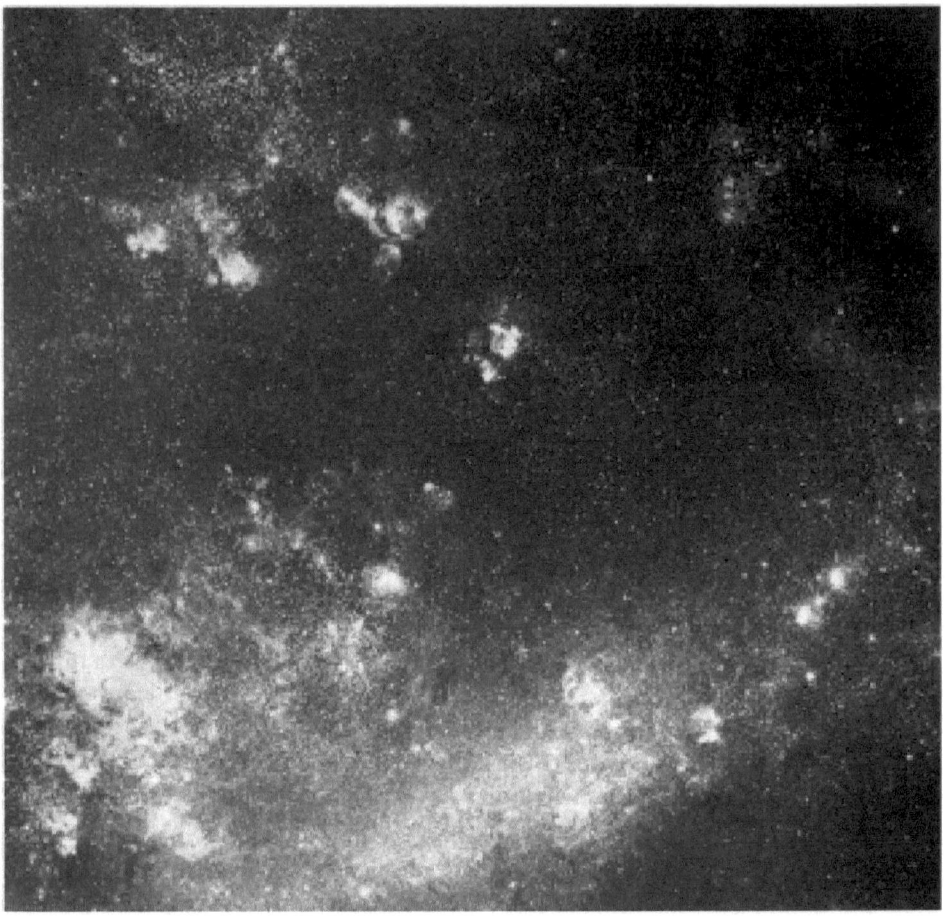

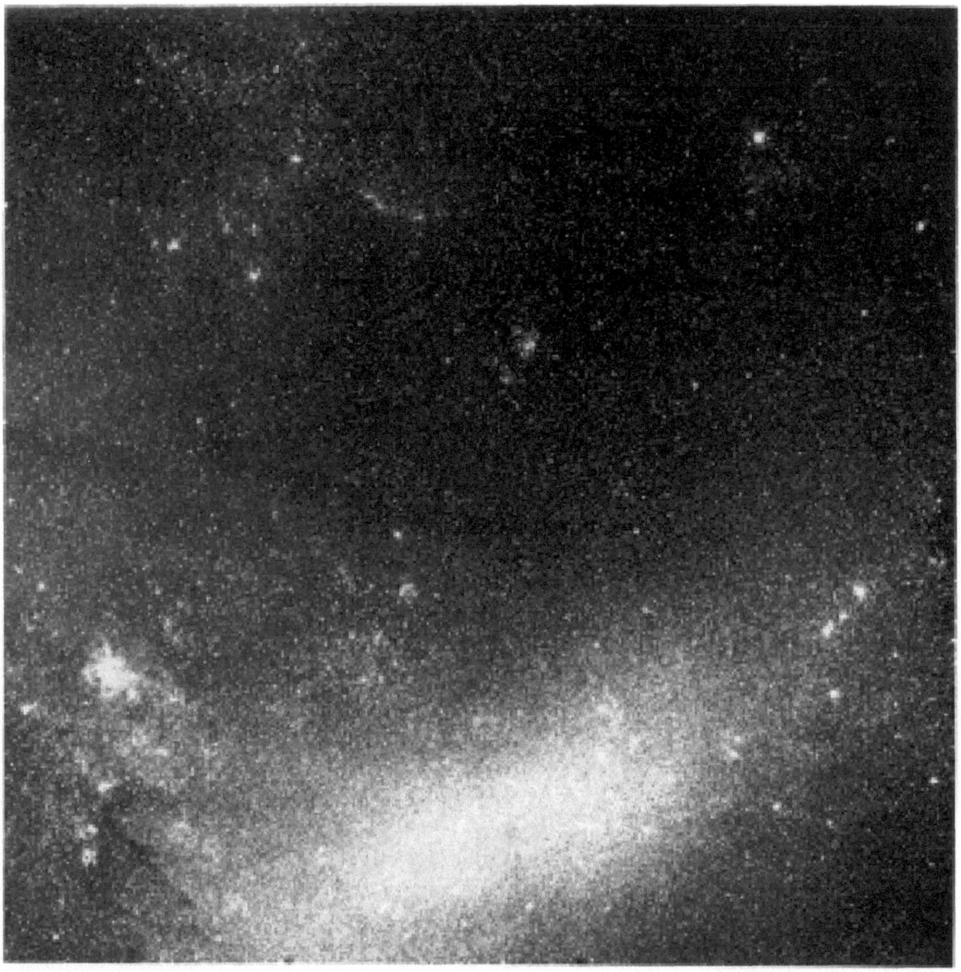

The Large Magellanic Cloud photographed in ultraviolet light (left) and in infrared light (right), taken by B. E. Westerlund with the Uppsala Schmidt telescope at Mt. Stromlo Observatory.

TABLE OF CONTENTS

PART III

CURRENT THEORETICAL APPROACHES

PART IV

PROSPECTS FOR FUTURE RESEARCH

PART I

PRINCIPAL PROPERTIES OF THE MAGELLANIC CLOUDS

SURVEY OF PRINCIPAL CHARACTERISTICS OF THE MAGELLANIC CLOUDS

A. D. THACKERAY
Radcliffe Observatory, Pretoria, South Africa

To the naked eye the two Magellanic Clouds look like detached portions of the Milky Way, well out of the galactic plane, at galactic latitudes 33° and 48°. One of the many important contributions to their study by the Australian radio astronomers has been to show that they are connected by a bridge of neutral hydrogen, although the resolution of the Parkes telescope has shown this bridge to be broken into stepping stones. Equally important was the discovery that the hydrogen density falls sharply at the boundary of the Large Cloud nearer to the galaxy. Thus the radio astronomers do not confirm the connecting bridge with the Milky Way that has been suspected by some astronomers. Viewed from outside the galaxy the Clouds would appear to be satellites situated 5 to 6 times as far from the galactic nucleus as the sun, and with diameters rather less than half that distance.

The currently accepted distances of the Clouds from the sun – 50 kpc for the Large Cloud and 61 kpc for the Small Clouds – can no doubt be refined with the help of large telescopes, but are not likely to be subject to any further major revision. It is accepted that they are at about one tenth the distance to the Andromeda nebula. Thus, as is well known, examination of the Magellanic Clouds with a 100-inch telescope is equivalent to examining the Andromeda nebula with a 1000-inch telescope. Since the Clouds can only be seen by those who live in the Southern hemisphere or who, like Magellan or astronauts, come south, the importance to cosmology and to all branches of astronomy of building large telescopes in the southern hemisphere is obvious. It is fitting that they should be the topic of this Symposium on the occasion of the dedication of the European Southern Observatory.

Both Clouds appear far more irregular than the average galaxy. Both have an elongated bar without any dominant nucleus. They serve as proto-types for the irregular Magellanic-type galaxies. The Small Cloud has an elongated wing pointing towards the Large Cloud which contrasts with the very smooth distribution of faint stars that characterises its main bar.

This wing is a relatively powerful source of neutral hydrogen 21 cm radiation and it tails off into a region with striking similarities to a spiral arm. Westerlund has shown that this region is probably the youngest in the Small Cloud, as judged by its extreme population I characteristics. As contrasted with the patchiness of the conspicuous population I characteristics, both Clouds probably possess population II haloes with a much more uniform distribution of faint stars.

As regards radio work in the LMC it is important to note the good spatial correlation between the radio 21 cm complexes and optical H II regions. We await news of

Muller (ed.), The Magellanic Clouds, 3–8.

radio surveys at the wavelengths of the higher H recombination lines. A negative search for OH has been reported.

Despite the general irregularities there is some support for de Vaucouleurs' classing the Magellanic Clouds as closely related to the barred spirals with ill-defined arms emanating from the ends of the bar. The detailed kinematics of such structures that can be studied with large telescopes form a problem of first-class importance.

Visvanathan's observations of polarisation of 30 LMC stars show the electric vectors to be uniformly aligned over large distances with directions apparently related to Cloud structure. For two stars within the bar the vectors lie along the bar where Mathewson has found a strong non-thermal component of radio continuum.

In the early days of Baade's concept of two populations, it was clear that the Large Cloud was very rich in population I with the enormous looped nebula 30 Doradus as its densest concentration. Baade even went so far as to suppose that the Large Cloud was pure population I. When RR Lyrae variables were first discovered in the SMC at the Radcliffe Observatory, some thought that the SMC might be pure population II. These were extreme views and it is now accepted that both Clouds are mixtures of the two populations or, as we would now prefer to say, include objects with a very wide range of ages.

The wide range of ages is probably best exhibited by the globular clusters. There are red globular clusters like NGC 121, 1466, 2257 containing RR Lyrae variables with colour-magnitude arrays by Gascoigne that closely resemble the halo globulars in our galaxy and are presumably equally old. There are blue globular clusters like NGC 1866, containing 3-day cepheid variables, probably less than 10^8 yr old according to Arp. Gascoigne finds others in the LMC with a weak or no red-giant branch but many faint blue stars. Gascoigne finds another group of globulars in the Small Cloud without a blue horizontal branch, like NGC 2158 and probably of intermediate age. Clearly there is an enormous amount to be learnt about the evolution of Cloud globulars which tax the powers of existing telescopes and observers to the very limit.

Our knowledge of the numerous open clusters in both Clouds is really restricted to what we would call associations in our Galaxy. The Pleiades for instance would appear only about 20″ of arc in diameter at the distance of the Clouds. Colour-magnitude arrays show a strong vertical blue branch as in young galactic clusters. Red giants and supergiants are also found in smaller numbers. Stars of intermediate colours are now attributed mainly or entirely to the galactic foreground. The Herstmonceux group found evidence for this from proper motions. The Hertzsprung gap is a feature common to these clusters in Clouds and Galaxy.

By contrast, in the general field of the LMC some super-supergiants are found scattered right across the gap. These bright stars in the Clouds offer a unique opportunity to study the evolution of massive stars. Moreover, at the distance of the Clouds a cross-motion of 10 km/sec will shift a star only 36″ of arc in a million years. Thus we can pinpoint the birthplace of a star less than 10^7 yr old with remarkable accuracy. We now have a clear picture of the upper limit to a star's luminosity – rather less than a million Suns. Bolometrically, the upper limit is about constant from type O to G5. Within the galaxy we have been almost powerless to establish the existence of such luminous stars.

This leads to the topic of calibration of luminosities of *all* types of objects that can be observed within the Clouds. Apparent magnitudes are a direct measure of luminosity, apart from uncertainties due to absorption and depth within the Clouds. The foreground visual absorption according to Radcliffe work amounts to only about 0.2 magnitude; this is the same as found for Cloud stars outside H II regions, and about half that for bright Cloud stars within H II regions. The opportunity to compare luminosities of supergiants, giants, main sequence stars of all types, Wolf-Rayet stars, cepheids, eclipsing variables, Mira-variables, RR Lyrae variables, globular clusters, novae, and so on in an endless sequence all at the same distance simply never occurs in our Galaxy.

We are only at the beginning of the task of identifying within the Clouds individual instances of some of these objects familiar to us in the galaxy. For instance although Westerlund's objective prism survey has shown statistically the presence of M supergiants in the Large Cloud, I think I am right in saying that no single case has been proved to be a Cloud member with slit spectra. We shall hear later about a current search for Mira type variables in the Clouds.

By far the most important application of the direct observation of luminosities within the Clouds has been in the establishment of the Period-luminosity Law for cepheids. We shall be hearing from the Gaposchkins of the vast amount of observation and reduction devoted at the Harvard Observatory to this corner-stone in the distance scale of the Universe. The work of Arp, Gascoigne and the Herstmonceux group has yet to be fully reconciled. Further photoelectric work at Boyden Observatory (Wayman-Butler) and Radcliffe Observatory (P. Andrews) seems to be helping to do this. But in any case, much more work on cepheids scattered over both Clouds will gradually elucidate the problem of an intrinsic spread in the P-L relation and add deeply to our ideas of the phenomenon of cepheid pulsation.

One class of Cloud object, well within the reach of small telescopes that has been very poorly studied is the ordinary nova. Novae probably occur about once per annum in the Clouds. Since 74-inch telescopes have been operating in the southern hemisphere, two Cloud novae have been discovered, one of which was announced a year after its photography.

An organised search for Cloud novae is urgently needed, with close intercommunication between centres in South America, Australia and South Africa.

1. Kinematics

A. ROTATION AND MASS OF LMC

The Sydney observations of 21 cm radiation from the LMC first demonstrated that the Cloud was subject to differential rotation. The optical radial velocities of supergiants observed at the Radcliffe Observatory indicated much more rapid rotation, but revised interpretation of the 21 cm profiles brought good agreement. It is now accepted that the major axis for rotation lies in p.a. about 171° (in agreement with the distribution of OB supergiants and some other objects) and that the inner core of the Cloud rotates like a solid body with a period of about 2.5×10^8 yr depending on the assumed inclination of the system, i.e. almost the same period as the sun's revolution round the Galaxy. The velocity dispersion of OB stars appears to be as low as 10 km/sec as in the galaxy, and this compared with the rapid rotation strongly suggests a flattened system, as originally proposed by de Vaucouleurs from structural features. The inclination of the system to the plane of the sky is still subject to great uncertainty but is probably near 30°. According to the latest estimate by Feast, the total mass of the main body (proportional to $\text{cosec}^3 i$) is about $5 \times 10^9 \odot$, probably less than 3% of the mass of the Galaxy. Neutral hydrogen accounts for about 10% of the total mass.

One of the striking features of the observed differential rotation is that the *centre*, appears to lie 1° north of the main bar, that is near the centre of gravity of the neutral hydrogen. This result applies to both OB stars and gas. A similar situation seems to occur in some other galaxies.

The only kinematic data on older population objects so far available comes from Feast's work on planetaries. The velocity dispersion is about twice as great as that for population I, but the rotation curve appears very similar, although Miss L. Webster's measures suggested a centre nearer to the optical bar.

B. SMC

Despite the more uniform distribution of OB and faint stars through the main body of the SMC, the kinematic picture is extraordinarily confused. Hindman has published a rotation curve based on median velocities with major axis in p.a. 55° and a gradient of over 20 km/sec per degree. Optical radial velocities in the SMC are still scarce. Unpublished Radcliffe velocities of improved quality give some support to Hindman's local mass motions, but do not as yet suggest any *general* rotation as rapid as that found by Hindman.

Hindman unexpectedly discovered numerous 21 cm profiles with double peaks extending over wide regions of the SMC. He found some support for the concept of expanding gas from the few cases of interstellar K recorded in Radcliffe spectra of Cloud numbers. Feast has found that his SMC planetaries show a double peak in the distribution of radial velocities. According to Hindman the confused dynamical picture of the SMC is due to expanding supernova remmants. On the other hand the cases of supernova remnants suggested by both radio (Mathewson) and optical (Westerlund) observations all occur in the Large Cloud.

2. Chemical Compositions

The brightest stars in the Clouds can be studied with sufficiently high spectroscopic dispersion for curve-of-growth analyses. We can also compare abundances of elements in the Cloud nebulae with those in the Galaxy. The first MK classifications of spectra in both Clouds by the Radcliffe team failed to show any sign of a *general* difference between the Clouds and the Galaxy. Przylbyski's analysis of the brightest star HD 33579 failed to show any striking peculiarity, though there was a faint suggestion of the ratio log (Fe/H) being 0.2 smaller than the solar value. It is conceivable that such marginal differences could be due to an expanding shell known to be present around such supergiants. Aller and colleagues have found that He, O and Ne are some 20% deficient relative to H in the Clouds.

Despite these almost negligible spectroscopic differences, there has always been a suspicion that the Small Cloud, in particular, differs in some fundamental way from the Large Cloud and Galaxy because of the distribution of periods of cepheid variables. The Small Cloud contains an abnormally high proportion of cepheids with periods of order 2 days, and abnormally large amplitudes. Such variables are rare in the galaxy and one galactic example TV Cam was believed to have an abnormal composition; but the spectroscopic peculiarity is now attributed to micro-turbulence. Even so the predominance of 2-day cepheids in the SMC (many of which *do* have the normal small light-amplitudes) very probably points to some difference from LMC and galaxy in its stage of evolutionary development.

One spectroscopic difference which perhaps points to the same conclusion is Feast's study of the excitation classes of the planetaries. The SMC planetaries are predominantly of low excitation while the LMC contains a high proportion of medium and high excitation. The whole picture of the LMC suggests to us that the system is one of very active star formation, while in the SMC apart from the wing there appears to have been a decline in the rate of star-formation. Bok and colleagues have found very rich LMC associations, embedded in clouds of hydrogen sufficient to form millions of suns in which the red supergiants appear to be confined to the outer edges. The belief is that the whole association is expanding and that the red supergiants have had time to evolve to that state during the expansion.

One vexed question concerns the dust content of the Small Cloud. The cepheids and OB supergiants that have been studied are surprisingly blue. Yet *some* dust is found in both Clouds as evidenced by reddening and especially Hutchings' detection of 4430 absorption. More work is needed and is in progress on this topic.

In conclusion it may be useful to summarise some of the salient features and overwhelming opportunities offered by the Clouds:

1. One tenth the distance of M 31.
2. All objects at essentially same distance.
3. Little interference from galactic absorption.
4. Young stars seen very near their birthplace.
5. Richness in cepheids provides best test for pulsational theory.
6. Opportunities for detailed regional studies of radio sources, thermal and non-thermal and comparison with optical sources.
7. Comparison of atomic abundances in galaxy and external system.

8. Spatial distribution of objects of various types within a system.

9. Determination of upper limit to a star's luminosity, with consequent pointers to theories of evolution of massive stars.

10. Studies of regional magnetic fields and polarisation in a system.

11. Brightest stars can be used as an external frame of reference for proper motions within the galaxy.

COLOUR-MAGNITUDE ARRAYS OF THE BRIGHTEST STARS

A. D. THACKERAY
Radcliffe Observatory, Pretoria, South Africa

Slit spectroscopy in the region of the Magellanic Clouds has the satisfaction of providing an immediate answer to the question whether any particular star belongs to the Cloud or to the galactic foreground. Examination with a hand eyepiece can usually give the answer quite conclusively; the star is a Cloud member if the spectrum shows.

(a) the sharp lines characteristic of high luminosity;

(b) the large velocity shift (about +270 km/sec in LMC or +160 km/sec in SMC) compared with most galactic stars.

Fehrenbach's objective prism technique has rested primarily on criterion (b) but this is not always sufficient by itself. Some high-velocity stars belonging to the population II galactic foreground have crept into his lists.

Before the Fehrenbach survey of the LMC we had the HDE classifications covering the LMC but *not* the SMC as a guide to picking up Cloud members through slit spectroscopy. Almost without exception a star with HDE classification B or 'Con' in the region of the LMC is an early-type Cloud member. However, quite early in the Radcliffe work Feast and Thackeray [1] reported the discovery of a few 'super-supergiants' of type F or G, visually among the brightest members of the LMC. Fehrenbach (2) has found more and this contribution is largely concerned with further Radcliffe work, spectroscopic and photometric, on these interesting supergiants of type A 0 or later which are probably in a stage of rapid evolutionary change.

We have reason to suppose that the census of Cloud members is now nearly complete through the uppermost two magnitudes. Such stars form the most reliable observational guide we have to theories on the evolution of the brightest and most massive stars.

Table I presents all stars of type A 0 or later known to be LMC members from slit spectroscopy. There are 62 stars in all, only 26 of which (including 7 variables) appear in the 1961 Radcliffe list [3]. The majority of the additional 36 stars were discovered by Fehrenbach's group [4] but some were discovered by the writer [5].

P. J. Andrews has measured UBV photometrically for many of the new stars at the Cape Observatory and his measured values of V are included in Table I. Figure 1 shows the 3-colour plot of Andrews' measures together with some of Wesselink's [6] UBV measures with the Radcliffe reflector. Johnson's UBV standards among supergiants do not extend later than B 9, but these agree well with the LMC early-type supergiants (assuming a small amount of reddening). Johnson's UBV for the main-sequence is also shown with the well-known dip at A 0; the dip practically disappears for the Cloud supergiants because the Balmer jump is so weak [7].

Figure 2 shows the HR diagram for the brightest stars in the LMC (omitting W and

Muller (ed.), The Magellanic Clouds, 9–18. *All Rights Reserved.*

TABLE I

LMC stars A 0 and later

HD	R	Sp	m_{pg}	Remarks
33579	76	A 3 Ia-0	9.29	
35343	88	A eq	var	S Dor
268668		A 2 (II:)	13.2[a]	
268675	61	A 0 Ia	10.85	
268757	59	G 5 Ia	11.8:	variable
268835	66	A eq	10.79	
268993	72	A 0 Ia	12.03	
268946	75	A 0 Ia	10.39	
269154		F 5	10.76	double; Mich. 904
269172	80	A 0 Ia-0	10.70	
269183		A 0: Ib: (e)	11.64	
269187		A 3 Ib	11.6[a]	
269236		F 2: Ia	11.5[a]	
269355		F 5 Ia	11.65	
269541		A 3 Ia (e)	11.2[a]	Westerlund 1962/26 CPD −68° 372
269563		A 2	12.2[a]	
269566		A 5 Ia	11.06	
269594		F 8 Ia	11.19	
269612		F 2 Ia	11.68:	
269628		A 0:	12.3[a]	
269634	106	A 0 I	12.6:	
269638		A 0:	12.0[a]	He I may be present
269678		A 2:	12.3[a]	
269682		A 0:	12.2[a]	
269694			12.3:	
269697		F 5 Ia	10.72	
269708			12.35	
269723	117	G 0 Ia	10.95	
269728		A 5:	12.4[a]	
269735		A 5 I (b)	11.91	
269781	118	A 0 Iae	9.94	
269787	119	A 0 Ia-0	11.3[a]	
269788		K 3: I	11.44	
269807		A 5 Ia	10.8[a]	
269809		A 5 I (b)	12.08	
269840		A 8:	10.69	
269841		A 0	11.9[a]	
269857		A 5 Ia	10.63	
269860		A 5 I (b)	11.90	
269905		–	11.7[a]	
269953	150	G 0 Ia	10.76	
269982	151	A 5 Ia:	11.4[a]	
269998		A 5:	12.10	
270033		A 8:	11.9[a]	
270046		G 0:	11.13	
270050		F 5 Ia	11.7[a]	
270086	153	A 1 Ia-0	10.52	
270111		F 8 Ia	10.88	
270402		A 0	12.20	
271182	92	F 8 Ia	10.24	
271192	98	A 0 Ia-0:	10.65	

Table I (continued)

HD	R	Sp	m_{pg}	Remarks
271 279	104	A 0 Ia:	11.34	
271 369		A 1	11.49	
W Men	102	F 8: Ip	var	
HV 873	60	F 8: I	var	cepheid
HV 5497	63	G 2 I	var	cepheid
HV 883	69	G 2: I	var	cepheid
HV 2294	77	F 7 I	var	cepheid
HV 2369	83	F-G	var	cepheid
HV 2447	91	G 0 Ia	var	cepheid
Anon	143	F 7 Ia	11.36:	Westerlund 30 Dor/15 in nebulosity; Westerlund magnitude.
Anon		A 8:		$5^h44^m.1$, 68°22 (1875). Fehrenbach misidentifies with HV 6029

[a] m_{pg} derived from the magnitude quoted in H.A. 100, no. 6, corrected by +0.3 as suggested by direct photometry of other bright Cloud members.

All other magnitudes in this column are photoelectrically observed B by P. J. Andrews (hitherto unpublished) or by Wesselink (*Monthly Notices Roy. Astron. Soc.* **121**, (1960) 337, or for R 143 by Westerlund (*Uppsala Ann.* **5** (1961), no. 1.

Pec stars) with V_0 (including absorption correction) plotted against spectral type. The upper envelope of stars in the top left corner corresponds to constant bolometric magnitude, as before. The right-hand side of the diagram is far more complete now, thanks to the work of the Fehrenbach group. In fact it seems doubtful whether additional Cloud members will be discovered with $V_0 < 11.0$ except perhaps for a few heavily reddened ones. The diagram shows a dozen super-supergiants later than F 0 and about a magnitude or more brighter than the brightest cepheids.

The Small Cloud has taxed the limits of objective prism equipment and for the most part slit spectroscopists have had to grope laboriously to find SMC members other than the 11 early-type members included in the HD catalogue. Last year the position was vastly improved by the appearance of Sanduleak's [8] finding list based on objective prism spectra with the Michigan Schmidt. This list includes 169 objects of which less than one quarter were previously known to be SMC members. Every single new case investigated by Radcliffe slit spectroscopy has so far proved to be a Cloud member. Sanduleak's criterion for membership is solely one of luminosity (criterion (a) above) and it is believed that this success springs largely from the fact that the spectra cover the Balmer limit with good definition.

A crude HR diagram for the brightest SMC stars has been built up in Figure 3 for comparison with Figure 2 in the following way. Open squares represent stars in the Radcliffe 1961 list. Wesselink's photometry and Radcliffe classifications being used to derive corrections for absorption 3.0 E_{B-V}; 3 cepheids with Arp photometry and Feast's spectral classifications also appear as open circles. The remaining plots are all based on Sanduleak's classifications and photographic magnitudes. Triangles re-

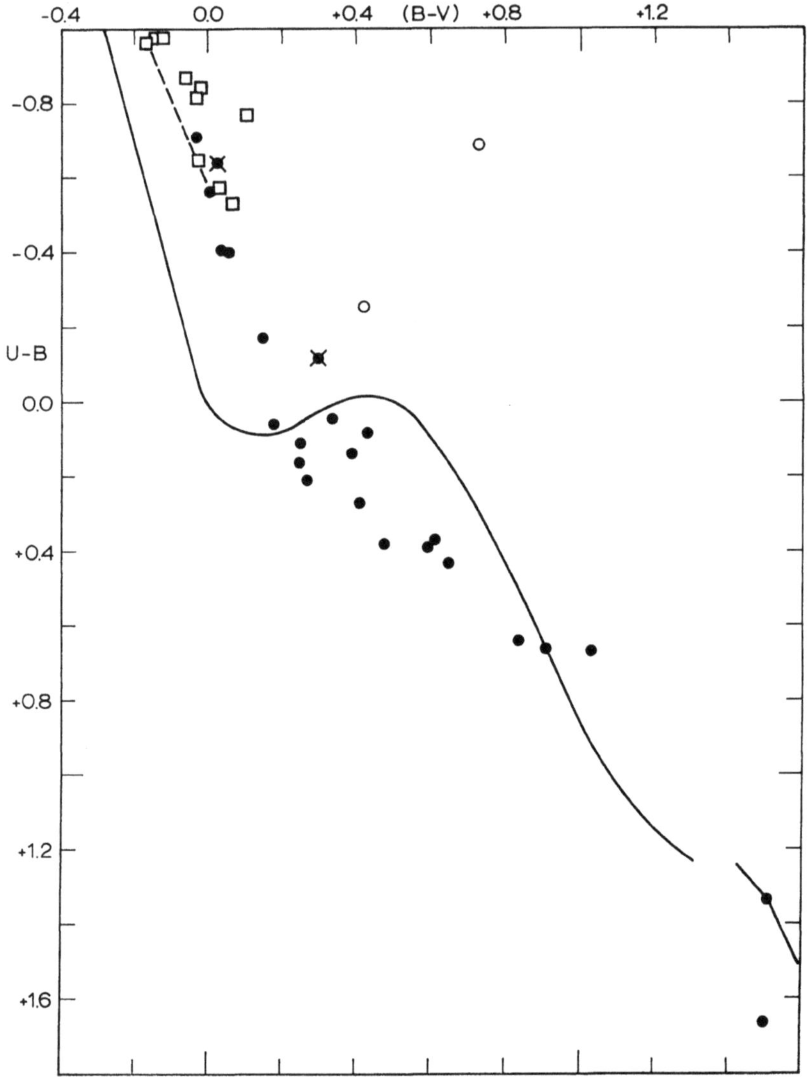

Fig. 1. Three-colour diagram of proved members of LMC. Filled circles – Andrews (Radcliffe); Open squares – Wesselink (Radcliffe); Open circles – double stars; Crossed circles – emission-line stars; Continuous curve – main-sequence; Dashed curve – galactic MK Class I

present previously known members without photometry. Sanduleak's m_{pg} has been transformed to V (ignoring absorption) for these remaining stars by means of a table of intrinsic colours $(B-V)_0$ for supergiants based on Radcliffe LMC work and Feinstein's galactic studies. These 'visual magnitudes' have been plotted against Sanduleak's spectral types. The large number of 'OB' stars have been dispersed about a mean of B 2.5 (as suggested by those for which slit classifications are available)

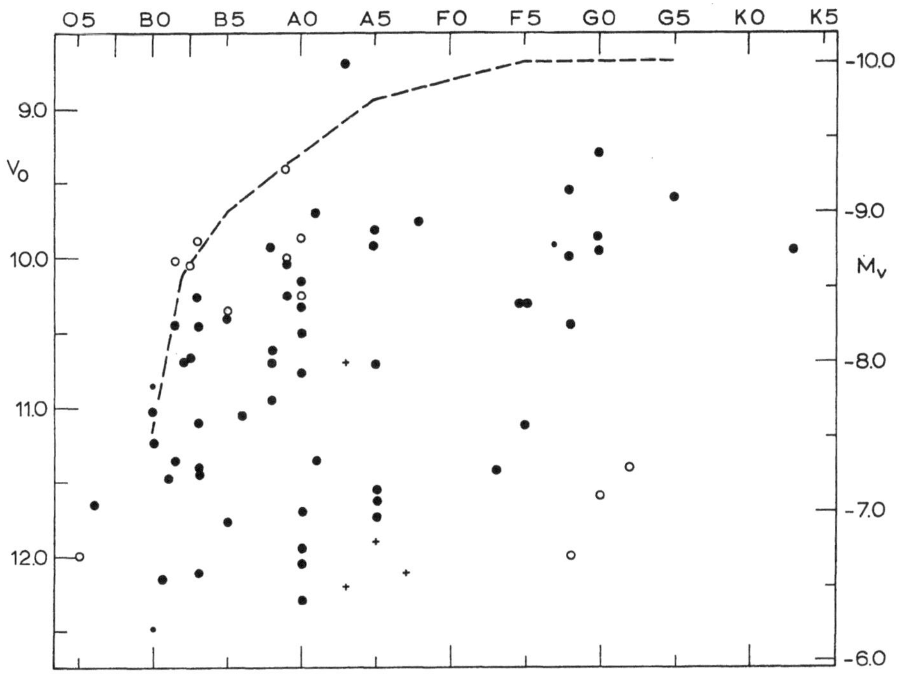

Fig. 2. LMC HR Diagram for proved members. Radcliffe spectra and V_0 derived from Radcliffe photometry (circles) or from HDE m_{pg} (crosses). Open circles (top left) are emission-line stars, (bottom right) are cepheids. S Dor, Pec and W stars omitted. The dashed curve corresponds to $M_b = -10.0$.

and the plots at A 0 and A 2 have been similarly dispersed. This should give a reasonable picture of the density of points in the HR diagram. The single point at F 5 is the important star S 118, for which Dr. Andrews has kindly communicated a rough visual magnitude of 11.0 as shown in the plot. The star, whose Cloud membership has been proved by R. Wood at the Radcliffe Observatory, has a close companion which renders photometry difficult.

This star S 118 and the variable K 5 star R 20 (=Henize S 30) are the only known stars appearing in the plot later than F 2, apart from the three cepheids.

The distribution of the triangles (previously known SMC members) shows that Sanduleak's finding list has added new members almost entirely restricted to the region below $V=12.0$. This fact is of considerable importance to the following discussion.

Comparison of Figures 2 and 3 shows that among the earliest types the distribution of points is essentially the same. Sanduleak's dense concentration of OB stars between $M_v = -6$ and -7 could no doubt be reproduced in the LMC were a suitable survey to be made. As was previously known, the SMC contains no counterpart to the exceptionally luminous ($M_v \sim -10$) LMC objects HD 33579, S Dor and HD 38268 (centre of 30 Dor); the latter two objects are excluded from Figure 2. But the really

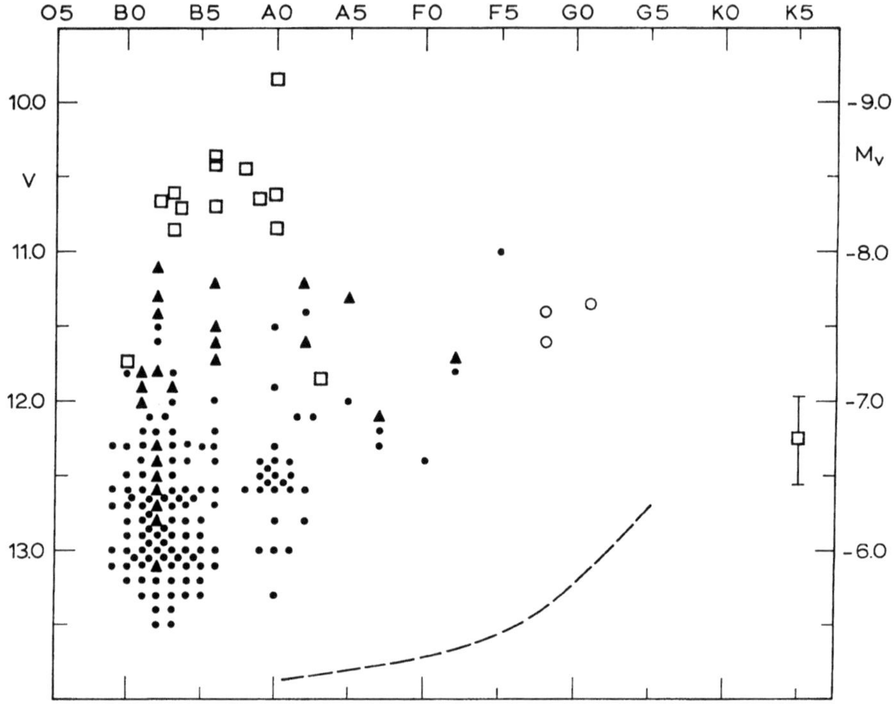

Fig. 3. SMC HR Diagram. Open squares – Proved members with Radcliffe photometry – Triangles – Proved members with Sanduleak m_{pg} converted to V – Closed circles – Sanduleak stars with m_{pg} converted to V – Open Circles – cepheids from *Monthly Notices Roy. Astron. Soc.* **121**, 337 – Dashed curve shows adopted $(B-V)_0$ as function of Sp.

striking difference occurs among the super-supergiants on the right of the diagram. Compared with the eleven LMC super-supergiants brighter than $M_v = -8$ and later than F 2, there is only the one known SMC object (Sanduleak 118) and even that appears to be fainter than any in the LMC group. Can there be other undiscovered super-supergiants in the SMC, lost through overcrowding or some other cause? It is possible, especially if the classification of *F-G* supergiants on the Michigan Schmidt spectra depends as much on the Balmer limit as the classification of OB does, for these red stars will necessarily be very weak in the ultra-violet. However, there are many good luminosity criteria for *F-G* supergiants at longer wavelengths, and it seems unlikely that *all* such objects with M_v brighter than -8.0 would have been missed had they been present in the SMC. As noted above, Sanduleak has in fact added few SMC members brighter than $M_v = -7$, and one would expect that some super-super-giants would have turned up in the search by Radcliffe spectroscopists had they existed in the SMC.

The position is summarised by the Hess diagrams in Figure 4. Here we present the percentage distributions of the various spectral classes as found in

(a) galactic Ia supergiants from the Jaschek Catalogue of MK classifications (252 stars).

(b) Large Cloud (194 stars).

(c) Small Cloud (169 Sanduleak stars).

(d) 1775 objects in the HDE Catalogue covering the Large Cloud (omitting clusters and nebulae).

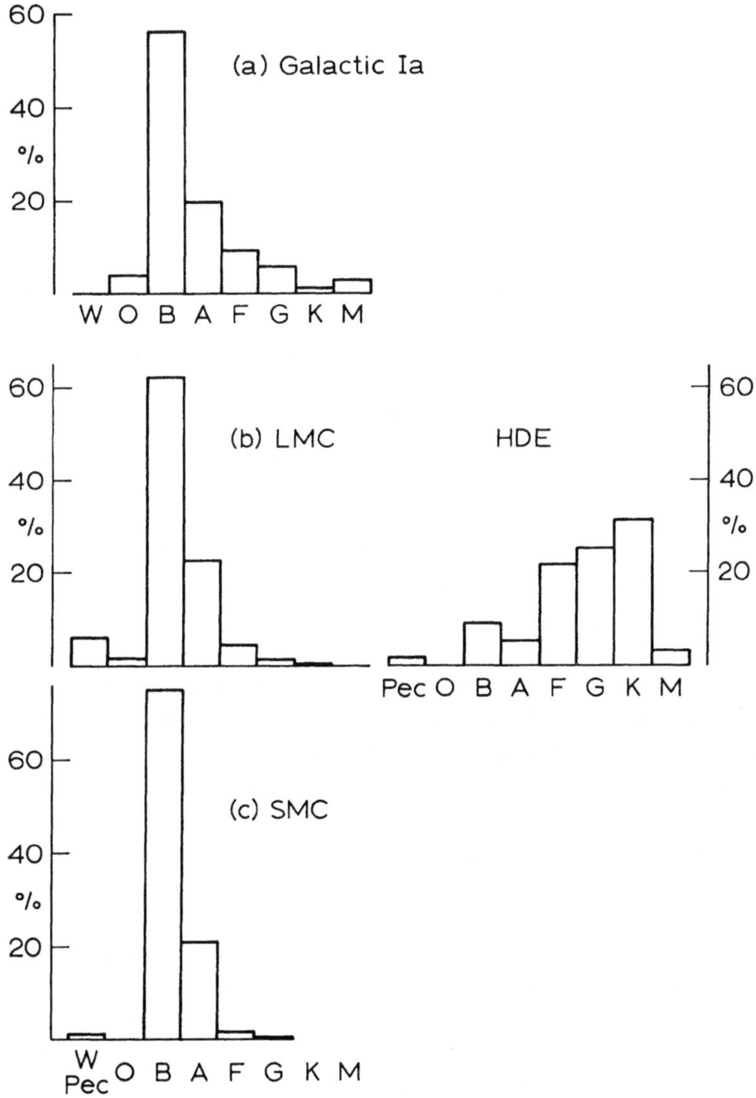

Fig. 4. Hess Diagrams of (a) Galactic Ia supergiants, (b) Proved LMC supergiants, (c) SMC supergiants (including Sanduleak stars).
The histogram at middle right shows the Hess diagram for HDE stars in the field of the LMC (mostly foreground).

It should be mentioned that slit spectra have *not* been obtained of all the 194 stars in group (b). It has been assumed that all stars classified in the HDE as B or Con (without slit spectra) are true LMC members and possess an identical Hess diagram to those which have been classified on the MK system.

Compare first (b) and (d). It is obvious that the vast majority of HDE stars classed as F to M are foreground galactic stars. On the other hand only one HDE 'B' star has been found to be foreground. The contrast between Figure 4 (b) and (d) emphasizes the difficulty of picking out the super-supergiants in the HDE and the value of the objective prism surveys of Fehrenbach and Sanduleak.

Comparing Figure 1 (a), (b), (c), the main differences between the LMC and the galactic Ia supergiants seem to be that

(i) we have not yet reached the LMC Ia supergiants of type M and it seems probable that their absolute magnitudes are fainter than the figure (−7:) provisionally allocated by Blaauw [9].

(ii) the LMC provides examples of W and Pec stars of high luminosity that are not recognised as class Ia in the galaxy. Otherwise the Hess diagrams for galactic Ia and LMC supergiants are remarkably similar.

(iii) Some galactic Ia *F-G* supergiants like FU Ori [10] may not be as luminous as supposed.

The SMC diagram stands out with its dominance of B stars and relative lack of the later types. Later than A 9 we have only 2.4% SMC members compared with 6.6% (LMC) and 20% (Galaxy). *If* the Sanduleak survey has not been affected by a selective difficulty in recognising the late-type SMC supergiants, then we have found some spectroscopic difference between the SMC, LMC and galaxy. This is *not* a suggestion of fundamental differences in chemical composition. We reiterate that Radcliffe spectra have never indicated any such fundamental difference. But it does begin to look as if there is a statistical suggestion of a difference in evolutionary development of the short-lived super-supergiants in SMC and LMC. The one other spectroscopic characteristic known to me which suggests a progressive difference SMC-LMC-Galaxy is Feast's [11] excitation classification of planetaries.

Where have these LMC super-supergiants evolved from in the HR diagram? Theoretical models of the evolution of massive stars suggest that the blue supergiants are burning helium after which they rapidly become red supergiants like those of type M in h and χ Per – the galactic cluster which has dominated much theoretical thinking.

The LMC c-m array suggests that there is a relatively stable region at spectral type *F-G*. Is it possible that these super-supergiants have not evolved from the main-sequence and helium-burning stages at all, but are at the very beginning of their lives? Some of them show extremely intimate associations with nebulosity (esp. Radcliffe nos. 59, 143, 150). In fact all but one of them are actually within, or extremely close to, the boundaries of Henize's H II regions (see Figure 5); the association of these objects with H II regions is considerably more marked than that of the spectral types B 5–A 8. It is possible that these LMC super-supergiants are extremely luminous examples of a

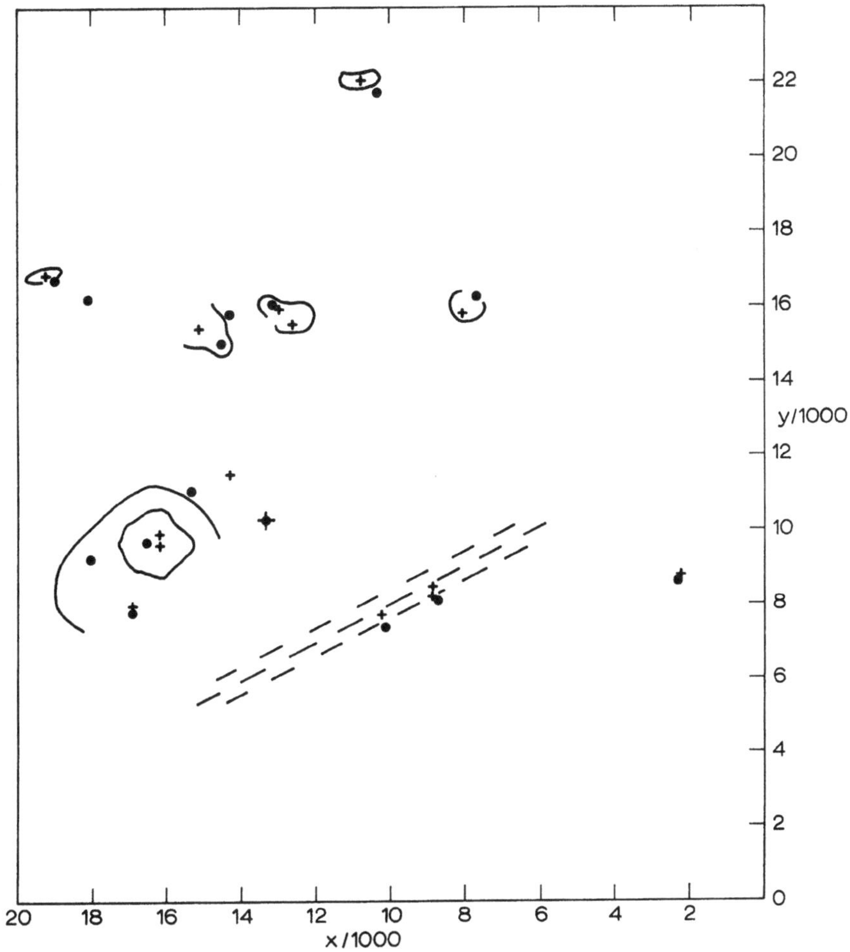

Fig. 5. Association of LMC super-supergiants (circles) with H II regions in plane of sky; Henize contours and point nuclei (crosses) are shown on Harvard x, y coordinates. Dashed lines indicate position of the LMC bar.

pre-main-sequence phase like that suggested by Herbig [10] for FU Orionis. If these objects are in fact concerned with the stage of star-formation, then we need only postulate that the LMC exhibits more signs of active star-formation than the SMC to explain the salient difference between the c-m arrays of the brightest stars in the two Clouds.

Subsequent to the writing of the above Martini [12] has communicated to the author his suggestion that the peculiar variable S Dor may have to be considered to be a pre-main sequence contracting star.

References

[1] Feast, M. W. and Thackeray, A. D.: 1956, *Monthly Notices Roy. Astron. Soc.* **116**, 587.
[2] Fehrenbach, Ch. and Duflot, M.: 1962, *Compt. Rend.* **253**, 1380. ESO Communication 1.
[3] Feast, M. W., Thackeray, A. D., and Wesselink, A. J.: 1961, *Monthly Notices Roy. Astron. Soc.* **121**, 337.
[4] Fehrenbach, Ch. and Duflot, M. J.: 1964, J. des Observateurs ESO communication 3.
[5] Thackeray, A. D.: 1962, *Monthly Notices Astron. Soc. South Africa* **21**, 74.
[6] Wesselink, A. J.: 1962, *Monthly Notices Roy. Astron. Soc.* **124**, 359.
[7] Thackeray, A. D.: 1957, Pontif. Acad. Sci. Scripta Varia, 16, Stellar Populations. 201.
[8] Sanduleak, N.: 1968, *Astron. J.* **73**, 246.
[9] Blaauw, A.: 1963, in Kuiper, G. and Middlehurst, B. (eds.) *Stars and Stellar Systems* **3**, 401.
[10] Herbig, G. H.: 1966, *Vistas in Astronomy* **8**, 109.
[11] Feast, M. W.: 1968, *Monthly Notices Roy. Astron. Soc.* **140**, 366.
[12] Martini, A.: 1969, private communication.

COLOUR-MAGNITUDE DIAGRAMS OF FAINT STARS IN THE ASSOCIATIONS AND IN THE FIELD OF THE MAGELLANIC CLOUDS

BENGT E. WESTERLUND

European Southern Observatory, Santiago de Chile

It is convenient to describe the two Magellanic Clouds in terms of sub-systems (Westerlund, 1965). In the Small Magellanic Cloud (SMC) we have the bar, the disk, the halo and the wing and in the Large Cloud (LMC) the bar, the central system, the disk, and the halo. Tables I and II summarize the observational data for the various systems.

A paper dealing with the stellar content of the Magellanic Clouds has been submitted by the author to *Vistas in Astronomy* and will appear this year.

Therefore, the discussion here will be limited to some general comments on the various subsytems and to presenting some new data on the wing of the Small Cloud.

1. The Sub-Systems of the Small Cloud

A. THE BAR

Hindman and Sinclair (1965) have shown that there is a very strong concentration of neutral hydrogen in the southern part of the bar. This has been interpreted as a substantial bar lying practically in the line of sight. The remainder of the optical bar is then probably connected with one of Hindman's expanding H I shells (Hindman, 1967).

The stellar population of the possible end-on-bar has been studied by Basinski *et al.* (1967). Further analysis of their photometric data (Westerlund, 1970) shows a rich extreme population I, similar to the population in and near NGC 330 (Arp, 1959b). Furthermore, there is a large number of intermediate-age population I stars. Its most luminous members are found at $M_v = -5$ mag, $B-V = +0.2$ mag. So far no counterpart to this component has been found in either Cloud. It is not yet established if the bar has an appreciable amount of red giants of population II. Arp (1959b) showed by star counts that there are about 18 times as many red giant stars near NGC 419 as near NGC 330. This means that there are only 0.3 red giants per 20000 square seconds of arc in the bar region.

B. THE DISK SYSTEM

Arp's (1959a) investigation of NGC 458 and our work on a number of faint clusters near NGC 419 (Westerlund, 1964b) show that an old population I exists in the Small Cloud. No stars of this type have been found in the wing area nor in the region of NGC 121 preceding the bar. In the bar region the photometry has not yet reached sufficiently faint stars to reveal this type of population. However, it appears possible

Muller (ed.), The Magellanic Clouds, 19–24.

TABLE I
The sub-systems of the Small Magellanic Cloud

Dimensions	Bar	Wing	Central system	Halo
	2°.5 × 1°	6° × 1°	7° (possibly flat)	7° (spherical)
Observed content	(a) *Extreme Population I:* OB associations, young clusters, supergiants of class Ia, H II regions.	*Extreme Population I:* OB associations, H II regions, blue and yellow supergiants of class Ia and I-0 long-period cepheids.	(a) *Population I:* Old open clusters planetary nubulae.	*Population II:* Old globular clusters, RR Lyrae variables.
	(b) *Population I:* Supergiants of class Ib Cepheids		(b) *Population II:* Red giant stars, intermediate-age globular clusters.	
Limiting magnitude:	17 mag	20 mag	20 mag	20 mag

Note: One degree is approximately 1 kpc. Hindman's (1967) results from H I observations indicate that the optical Bar should possibly be considered as an end-on bar plus a superposed arc (or arcs) of extreme population I objects. The centre of the SMC has been assumed to lie near NCG 419, and the length of the wing has been measured from this point. The extent of the 'central' system is based on Gascoigne's (1966) classification of the intermediate-age globular clusters.

TABLE II

The sub-systems of the Large Magellanic Cloud

Dimensions	Bar	Central system	Disk	Halo
	3° × 1°	6° (flat system)	14° (flat)	24° (spherical)
Observed content	(a) *Extreme Population I:* OB associations and H II regions (superposed on bar and belonging to Central system?). (b) *Population II:* Red giants	*Extreme Population I:* OB associations, young clusters, supergiants of class Ia and I-0. H II regions and most of the neutral hydrogen.	*Population I:* Old open clusters, planetary nebulae, carbon stars.	*Population II:* Old globular clusters. LMC-type globular clusters.
Limiting magnitude:	$I = 13.5$ mag	$V = 16$ mag	$V = 17$ mag	$V = 20$ mag

Note: One degree is about 1 kpc. The centre of the LMC has been taken to coincide with the centroid of the system of the planetary nebulae, R.A. $= 5^h24^m$, Dec $= -69°.5$ (1950). The classification of the globular clusters of the LMC is difficult. Gascoigne (1966) recognizes one group of LMC clusters which have no galactic counterparts and one cluster, NGC 1783, which cannot be classified.

that this old population I may be limited to a relatively small portion of the SMC.

The distribution of the population II red giant stars can be seen in Table III. Obviously the extent of this type of population is very limited. Its centre of gravity appears to fall outside the bar, possibly at about 1^h10^m, $-72\overset{\circ}{.}2$.

Arp (1958a, 1961) has suggested that a large number of the stars in the colour range $+0.4 - B - V - +1.0$ with $V < 17$ mag in the region of the Cloud are physical members of the SMC. These stars would then represent a class of objects of which there are few representatives in our Galaxy. Photometry by the author of a number of these stars near NGC 419 and of several stars near NGC 602 shows that most of the stars are of galactic halo type; thus are *not* Cloud members.

TABLE III

Number of red giant stars of population II per 20000 square seconds of arc in the SMC

Region at	Number of stars	Distance from core	Distance from NGC 419	Reference
NGC 419	6	1°.5	0°	Arp (1958a)
NGC 458	6.5	2.0	1	Arp (1959a)
NGC 361	5	1.3	1.4	Arp (1958b)
NGC 456–65	1.5	1.6	0.6	Westerlund (1961b)
NGC 330	0.3	0.5	1.0	Arp (1959b)
NGC 121	0	2.5	3.5	Tifft (1963)
NGC 602	0	2.5	1.7	Westerlund (1964a)

C. THE HALO

Very little is known about the SMC halo except from Tifft's (1963) discussion of NGC 121 and its surrounding. We find it not yet proven that any of the field stars ascribed by Tifft to the SMC halo are really physical members of the SMC. With regard to the globular clusters in the SMC L 1, K 3, NGC 121 and L 113 may be halo-type clusters, different from for instance NGC 419. However, more accurate and complete photometry of the clusters is needed before definite conclusions may be drawn.

D. THE WING

The wing extends from about 1^h10^m, $-73.4°$ out to and including Shapley's wing, at 2^h15^m, $-74.5°$ (Shapley, 1940). The analysis of the region of NGC 602 (Westerlund, 1964a) to a limiting magnitude of $V = 20$ mag, and of the region of NGC 456 – 65 to $V = 19$ mag (Westerlund, 1961) led to the conclusion that the wing consists of extreme population I, only. Recently (Westerlund and Glaspey, 1971), a search for blue stars has been completed for an area of 16 square degrees extending from 1^h30^m to 2^h15^m in Right Ascension. About 2100 blue objects were identified; the limiting magnitude of the search was $V = 18.5$ mag. The stars form a slightly irregular band through the investigated field. In it are several concentrations and clusterlike formations but also some areas with a pronounced lack of blue stars. The borders of the band are well defined in all directions.

Three-colour photometry of all stars inside two areas of 0.1 square degrees each has been carried out to a limiting magnitude of $V=17$. The identification of the blue objects is confirmed by the measured colours. The remainder of the stars are galactic main-sequence and halo-type objects. This photometry, combined with a previous study of some bright supergiant stars (Westerlund *et al.*, 1963) has lead to the conclusion that we are dealing with one age group, only, in the wing. To this group belong certainly also the yellow supergiants identified by Florsch and Carozzi (1965). Together with the long-period cepheids in the area they complete the picture of an extreme population I in the wing.

2. The Sub-Systems of the Large Cloud

A. THE BAR

Our knowledge of the composition of the bar is extremely limited. A number of late type M-stars have been found with an infrared magnitude $I=13$ mag (Westerlund, 1960). If they are members of the LMC then they are population II red giants. In the bar region are also seen some early type M supergiants, some blue supergiants, a few associations and H II regions. They belong most likely not to the bar but are a superposed population.

B. THE CENTRAL SYSTEM

This system contains all the super-associations of bright blue stars and emission nebulae in the LMC as well as practically all the neutral hydrogen. The limiting magnitude for photometry is so far only $V=17$ mag. Photometry of several of the so-called 'young population clusters' (Hodge, 1961) has been carried out by Hodge, and photometry of other clusters, associations and selected field areas have been carried out by Woolley, Westerlund, Hodge, Bok and others. (For references see Westerlund, 1970). The most important conclusion from a comparison of the various colour magnitude arrays is that all the present super-associations are of nearly identical age (Westerlund and Smith, 1964). This result combined with the study of older objects indicates that the star formation in the LMC has occurred in bursts.

C. THE DISK POPULATION

Also for this system the information is very incomplete. Obvious members are the planetary nebulae, the carbon stars, and the faint open clusters. Representative for the latter group may be NGC 1866 (Arp, 1967) and its surrounding field stars. In addition to the old population I component there is an indication of a population II component in the red giant branch seen in some colour magnitude arrays. However, these red stars may belong to the halo.

D. THE HALO

The only objects known in the halo are the globular clusters. Gascoigne (1966) summarizes the present available information: of the existing 35 large red clusters,

6 have been studied, 2 (NGC 1466 and NGC 2257) appear to be ordinary globular clusters ressembling the halo-type clusters in our galaxy, and 1 (NGC 1783) is unclassified. The remaining three form a class by themselves. Gascoigne suggests that they might be very old objects. It is extremely difficult to describe this population at present and much more observational data are needed before the halo population of the LMC is accurately known.

E. CONCLUSIONS

From the analysis of the colour-magnitude diagrams we have seen that both Clouds have an extreme population I occupying mainly the more central regions. The wing of the SMC forms an exceptional feature with an extreme population I extending to over 7 degrees from the main body. We note that the extreme population I in the Large Cloud contains a large number of young populous clusters of which there are few in the Small Cloud. The older population I forms a disk structure which in both Clouds may extend outside the youngest system. The composition of the bar in the LMC is not very well known, but it appears to be dominated by red giant stars. The bar in the SMC has an extreme population I as well as an intermediate age population I. Both galaxies have probably a halo population similar to that in our Galaxy. Both galaxies have also globular clusters, many of which may be of intermediate age. So far few halo stars, except some variables, have been observed outside the halo clusters in either galaxy.

References

Arp, H. C.: 1958a, *Astron. J.* **63**, 273.
Arp, H. C.: 1958b, *Astron. J.* **63**, 487.
Arp, H. C.: 1959a, *Astron. J.* **64**, 175.
Arp, H. C.: 1959b, *Astron. J.* **64**, 254.
Arp, H. C.: 1961, *Science* **134**, 810.
Arp, H. C.: 1967, *Astrophys. J.* **149**, 91.
Basinski, J. M., Bok, B. J., and Bok, P. F.: 1967, *Monthly Notices Roy. Astron. Soc.* **137**, 55.
Florsch, A. and Carozzi, N.: 1965, *Compt. Rend.* **261**, 2837.
Gascoigne, S. C. B.: 1966, *Monthly Notices Roy. Astron. Soc.* **134**, 59.
Hindman, J. V.: 1967, *Australian J. Phys.* **20**, 147.
Hindman, J. V. and Sinclair, M. W.: 1965, *Mt. Stromlo Symposium on the Magellanic Clouds*, p. 76.
Hodge, P. W.: 1961, *Astrophys. J.* **133**, 413.
Shapley, H.: 1940, Harvard Bull. No. 914, p. 8.
Tifft, W. G.: 1963, *Monthly Notices Roy. Astron. Soc.* **125**, 199.
Westerlund, B. E.: 1960, *Uppsala Astron. Obs. Ann.* **4**, No. 7.
Westerlund, B. E.: 1961, *Uppsala Astron. Obs. Ann.* **5**, No. 2.
Westerlund, B. E.: 1964a, *Monthly Notices Roy. Astron. Soc.* **127**, 429.
Westerlund, B. E.: 1964b, *IAU/URSI Symp.* **20**, 342.
Westerlund, B. E.: 1965, *Mt. Stromlo Symposium on the Magellanic Clouds*, p. 40.
Westerlund, B. E.: 1970, in A. Beer (ed.) *Vistas Astron.* **12**, 335, Oxford.
Westerlund, B. E., Danziger, I. J., and Graham, J.: 1963, *Observatory* **83**, 74.
Westerlund, B. E. and Glaspey, J.: 1971, *Astron. Astrophys.*, in press.
Westerlund, B. E. and Smith, L. F.: 1964, *Monthly Notices Roy. Astron. Soc.* **128**, 311.

CLUSTERS IN THE MAGELLANIC CLOUDS

S. C. B. GASCOIGNE

Mount Stromlo and Siding Spring Observatories, Research School of Physical Sciences, The Australian National University, Canberra, Australia

Star clusters are extremely common in the Magellanic Clouds. About 120 are known in the SMC, and more than 1600 in the LMC. The true number there has been estimated by Hodge and Sexton (1966) to be about 6000, this total referring to clusters with brightest stars brighter than $M_B = 2$. The outermost recognised members of the Cloud system are clusters, Lindsay 1 on the western side being 30°, or at least 30 kpc, from NGC 2257 on the eastern side. Attempts have been made to fit the Cloud clusters into some sort of structural pattern, without success; so far, no analogue has emerged in the Clouds to the galactic system of a narrow disk containing the young clusters surrounded by a spheroidal distribution of old clusters. The nearest approach has been Lynga and Westerlund's (1963) result that the outer LMC clusters define a quite good ellipse on the plane of the sky. The ratio of the axes is about $\sqrt{2}$, consistent with a disk-like distribution seen from an angle of 45°, but this angle disagrees with the 27° deduced from other evidence and the problem rests in abeyance.

The easiest properties of a cluster to measure are the integrated colours and magnitudes. These are now available, in B and V, with some also in U, for about 80 clusters in the two Clouds (van den Bergh and Hagen, 1968; Gascoigne, 1969), and when plotted as in Figure 1 exhibit the familiar dichotomy into blue and red groups, very few clusters indeed having colours in the range $0.35 < (B-V) < 0.60$. This dichotomy is not difficult to understand. The blue clusters are young, with most of their light coming from unevolved B stars at the bright end of the main sequence; the red clusters are old, with only a faint main sequence and most of their light coming from red giants; while the middle-aged clusters, like the Hyades, are too faint to appear in Figure 1 ($B \simeq 16$ or 17), their evolved stars being still too massive to have developed degenerate cores. Colour is then an indicator of age. One may remark how few young clusters there are in the SMC; also the considerable distances of some young clusters from the main body of the LMC. Thus NGC 2214, which cannot be older than 10^7 yr, is 3° (3 kpc projected distance) from 30 Dor, and 5° from the centre of the LMC bar.

Infrared measurements have so far been made only for two clusters (Kron, 1961), and narrow-band measurements of the G-band for four (Kohler, 1965). Both these lines could be worth pursuing.

Colour-magnitude diagrams for young clusters have been measured principally by Arp (1959), Woolley (1960, 1961, 1963), Westerlund (1961) and Gascoigne (1969). We show two examples in Figure 2. Both of Arp's clusters (NGC 330 and 458, in SMC) contain supergiants of intermediate colour which have been difficult to explain and have given rise to some controversy (Feast, 1964). NGC 2100 is typical of a number in the LMC, with very red supergiants ($B - V \simeq 2.0$) reminiscent of those in h and χ

Muller (ed.), The Magellanic Clouds, 25–30.

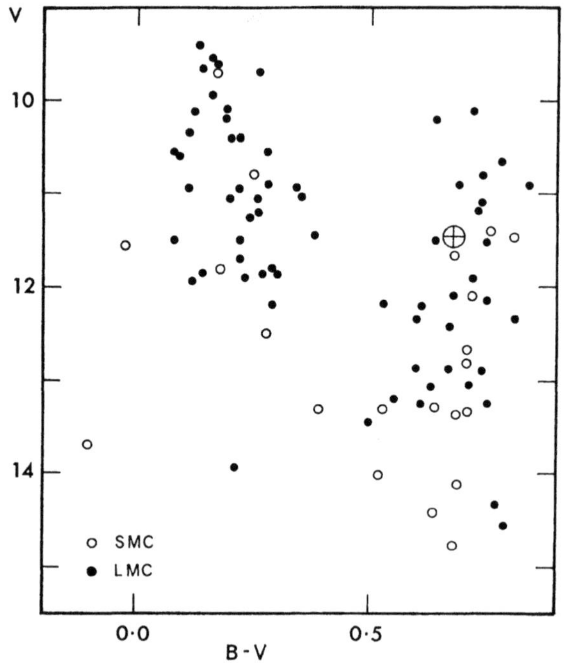

Fig. 1. Integrated colours and magnitudes for Magellanic Cloud clusters, from van den Bergh and Hagen, and Gascoigne. No reddening corrections have been applied. ⊕ indicates an average halo cluster seen at the distance of the Clouds.

Persei. An interesting result of Westerlund's is that WR stars tend to occur at the breakoff points of his clusters. These are all about the same age, $(3-10\times10^6$ yr) and appear to have the same initial luminosity function as clusters in the galaxy. But they have no known analogues in the galaxy, and we owe to Woolley (1961) the suggestion that they are young globular clusters. The LMC may thus throw some light on conditions in the galaxy at the time our own globular clusters were being formed.

Older blue clusters with colour-magnitude diagrams include NGC 1831 (Hodge, 1963), NGC 2031 (Gascoigne and Ford, 1969), and the five which contain cepheids (see Figure 3 and Table I).

TABLE I

LMC clusters with cepheids

	Integrated Magnitudes			
	V	$B-V$	n	Periods
NCG 1866	9.6	0.25	~ 10	$2^d.6-3.5$
NCG 2010	12.3	0.26	3	2.0, 2.8, 3.4?
NGC 2136	10.5	0.28	3	7.7, 9.5, ~ 10.8
SL 204			1	4.8
NGC 2031	10.9	0.28	3	3.0, 3.3, 3.3

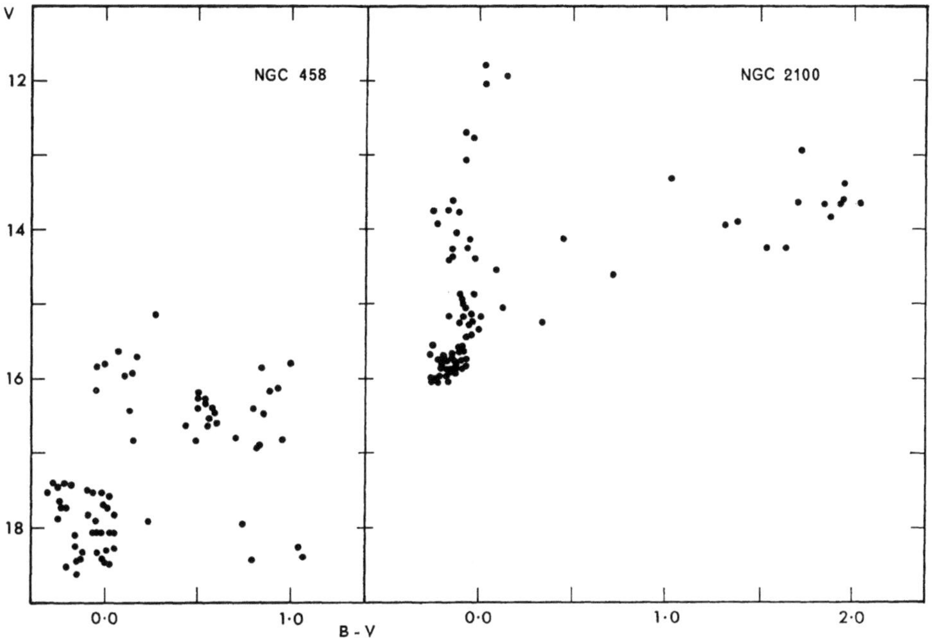

Fig. 2. Colour-magnitude diagrams for NGC 458 in the SMC (Arp) and NGC 2100 in the LMC (Westerlund). Westerlund's measurements did not extend below 16 mag.

Arp and Thackeray's (1967) work on NGC 1866 is most important, as providing direct confirmation that cepheids are a post red-giant stage in the evolutionary sequence (cf. Iben, 1967). NCG 2010 promises to be important also, in a different way. Hodge (1961) noticed that it contained two cepheids, to which they assigned periods, while Gascoigne and Hearnshaw (1969) found one more, and determined the following magnitudes as given in Table II.

HV 2593 is clearly too bright for its period and appears to be pulsating in the first harmonic. Its low amplitude and sinusoidal light-curve reinforce the similarity to c-type variables in globular clusters, and suggest moreover that the sub-class of cepheids isolated by Arp in the SMC, which are similarly too bright for their periods, and have low-amplitude sinusoidal light-curves, may be oscillating in the first harmonic.

TABLE II

	P	$\log P$	$\langle V \rangle$	$\langle V \rangle - \langle B \rangle$	V_{amp}
HV 2593	$1^d.968$	0.294	15.98	0.45	0.31
HV 2599	2.853	0.455	16.02	0.56	0.75
v 3	3.43	0.535	15.8		0.57

Generally an important attribute of these bright young clusters lies in the number of stars they contain, which may be sufficient to reduce statistical fluctuations to the point where reliable estimates can be made of the duration of the cepheid phase, and perhaps also allow empirical evolutionary tracks to be constructed in the lines of those of Sandage (1957) for M 3 and M 67. And if, as is now becoming possible, the photometry of individual stars of, say, NGC 458 could be pushed below $V=20$, we could use standard main-sequence fitting to determine a badly-needed independent modulus for the SMC.

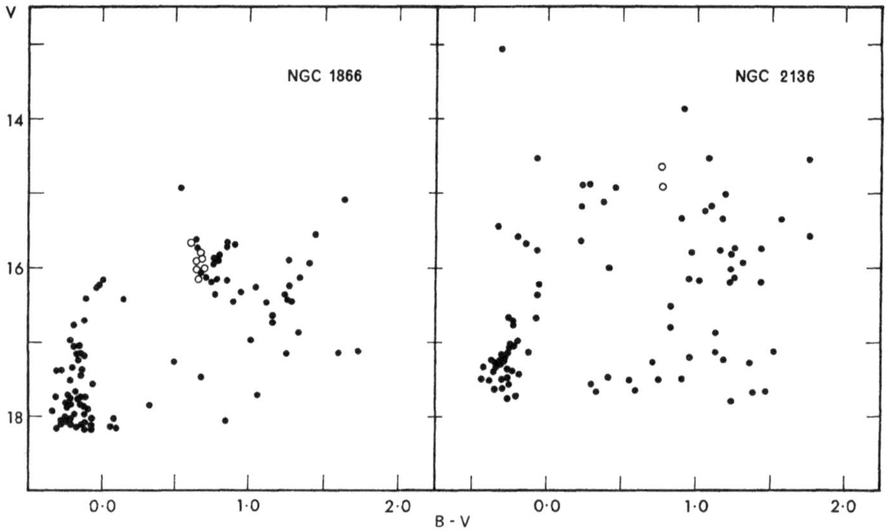

Fig. 3. Colour-magnitude diagrams for NGC 1866 (Arp and Thackeray) and for NGC 2136 (Gascoigne and Hearnshaw, preliminary). The open circles indicate cepheids.

The earliest and still the most significant observation of the old Cloud clusters was the discovery of RR Lyraes in four of them by Thackeray and Wesselink. In the 1950's this provided one of the strongest arguments for the then recent revision of the distance scale, and also established the existence of really old constituents in each Cloud. Subsequent work seems only to have complicated the positions. This has consisted mainly of the determination of colour-magnitude arrays for about 15 clusters, and has been well summarised by van den Bergh (1968, Table XVI). Three types of old clusters have been recognised in the Clouds.

(1) four globular clusters of traditional type, NGC 121 fairly near the SMC, the other three NGC 1466, 1844 and 2257, associated with but all at a considerable distance from the LMC.

(2) a type peculiar to the SMC, with no blue horizontal branch, but a well-populated giant branch, terminating in bright ($M_V \simeq -3$) very red ($B-V \simeq 1.9$) giants. These have been tentatively identified with the 'intermediate-age' clusters in the galaxy (Gascoigne, 1966).

(3) a type peculiar to the LMC, with very few red giants – hardly any brighter than $M_V = -1.5$ – and strong blue horizontal branches. These clusters appear to have no galactic counterparts. Figure 4 summarises the evidence. These clusters are difficult to observe, and further progress has had to await the development of means to measure fainter stars. The 'true' globular clusters have been very useful in providing independent moduli for the Clouds.

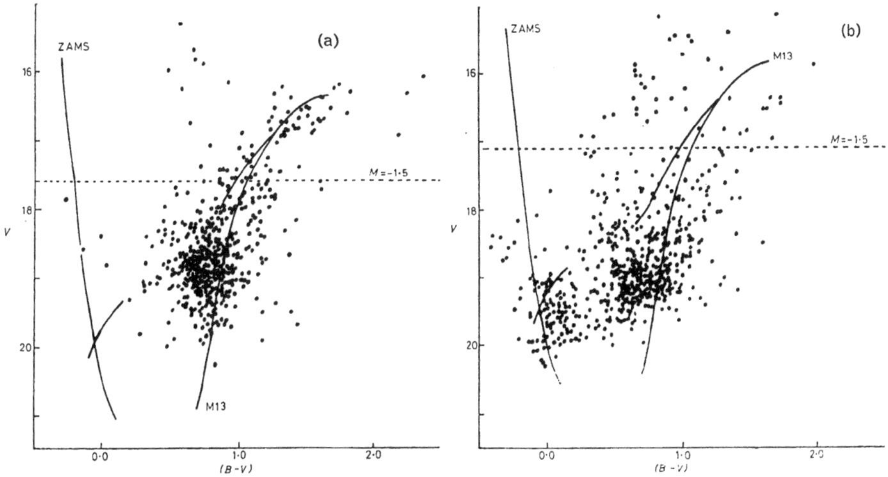

Fig. 4. Superposed colour-magnitude diagrams for (a) 3 SMC clusters and (b) 3 LMC clusters (Gascoigne, 1966).

Hodge and Wright (1963) have found population II variables in a number of red LMC clusters with periods from ~1 to ~13 days. Their estimates make these variables about a magnitude brighter than expected, an anomaly which requires attention. Two other points to mention are that Cloud clusters are often quite elliptical, indicating substantial rotational velocities; and their colour-magnitude diagrams are usually quite typical of those of the fields in which they are embedded, indicating a common time of origin.

It has long been recognised that radial velocities of the Cloud clusters have the potential to tell us a great deal about the dynamics of the Clouds – they are the most outlying identifiable members, and the analogy with the galaxy and M 31 is tempting. The clusters are however, faint and diffuse, and the exposure times required up to now have been prohibitive, although a few spectra had been obtained at Pretoria (Thackeray, 1959). With faster spectrographs, and especially the advent of the image tube, we may expect this picture to change. Integrated spectra can also, of course, convey much information about the clusters themselves.

A final remark also concerns a dynamical problem. Of the four 'true' globular clusters known to be associated with the Clouds, three at least are quite loosely bound.

One may ask, were they not originally members of the galaxy? More generally, how many galactic globular clusters would the Clouds capture (a) in a passing encounter (b) if they were permanent physical companions, and could either number be reconciled with the present four. The problem is analogous to the sweeping-up of asteroids by Jupiter. A converse question is also interesting. The LMC contains, in appreciable numbers, two types of cluster not known in the galaxy, rich young clusters like NGC 1818, 1866 and 2100, and giant-poor clusters of the type referred to above. One would expect a steady rate of loss of these clusters under the tidal force of the galaxy, in close analogy to the dissolution of galactic clusters. Can our failure to detect any of these clusters in the vicinity of the Sun be used to place an upper limit on the time during which the Clouds have been companions of the galaxy?

In the galaxy clusters have played a major role in establishing the distance scale, in providing much of the observational evidence, especially the time-scale, for stellar evolution, and in giving us our best line on the early history of the galaxy itself. In the Clouds their importance will be comparable. Already the clusters there have provided important data, especially on the evolution of massive stars, and have shown the existence of new phenomena concerning older stars. But the relation of clusters to processes of star formation, and to the structure and evolution of the Clouds, remains obscure, and may be expected to provide problems for many years to come.

References

Arp, H. C.: 1959, *Astron. J.* **64**, 175.
Arp, H. C. and Thackeray, A. D.: 1967, *Astrophys. J.* **149**, 73.
Feast, M. W.: 1964, in F. J. Kerr and A. W. Rodgers (eds.), 'The Galaxy and the Magellanic Clouds', *IAU Symp.* **20**, 330.
Gascoigne, S. C. B.: 1966, *Monthly Notices Roy. Astron. Soc.* **134**, 59.
Gascoigne, S. C. B.: 1969, to be published.
Gascoigne, S. C. B. and Ford, V.: 1969, to be published.
Gascoigne, S. C. B. and Hearnshaw, J.: 1969, *Proc. Astron. Soc. Australia* **1**, 208.
Hodge, P. W.: 1961, *Astrophys. J.* **133**, 413.
Hodge, P. W.: 1963, *Astrophys. J.* **137**, 1033.
Hodge, P. W. and Sexton, J. A.: 1966, *Astron. J.* **71**, 363.
Hodge, P. W. and Wright, F. W.: 1963, *Astrophys. J.* **138**, 366.
Kohler, J. A.: 1965, *Observatory* **85**, 197.
Kron, G. E.: 1961, *Publ. Astron. Soc. Pacific* **73**, 202.
Lynga, G. and Westerlund, B. E.: 1963, *Monthly Notices Roy. Astron. Soc.* **127**, 31.
Sandage, A.: 1957, *Astrophys. J.* **126**, 326.
Thackeray, A. D.: 1959, *Astron. J.* **64**, 437.
van den Bergh, S.: 1968, *J. Roy. Astron. Soc. Can.* **62**, 145, 219.
van den Bergh, S. and Hagen, G. L.: 1968, *Astron. J.* **73**, S206.
Westerlund, B. E.: 1961, *Uppsala Astron. Obs. Ann.* **5**, No. 1.
Woolley, R. v. d. R.: 1960, *Monthly Notices Roy. Astron. Soc.* **120**, 214.
Woolley, R. v. d. R.: 1961, *Proc. Roy. Soc. (A)* **260**, 189.

SUPERNOVA REMNANTS, PLANETARY NEBULAE AND RED STARS IN THE MAGELLANIC CLOUDS

BENGT E. WESTERLUND
European Southern Observatory, Santiago de Chile

Only brief summaries of our present knowledge of the three groups of objects referred to in the title will be given here together with references to the major papers containing the observational data.

1. Supernova remnants

Investigations with the 210-ft radiotelescope at Parkes have shown that four non-thermal radio sources in the Large Magellanic Cloud (LMC) coincide in position with the emission nebulae N 49, N 63 A, N 132 D and N 157 (30 Doradus) in Henize's catalogue (1956). It was suggested that on radio wavelengths the 30 Doradus region is very similar to the central region of our Galaxy, and that the three other radio sources may be supernova remnants (Mathewson and Healey, 1964). A comparison of the radio and the optical properties of N 49, N 63 A and N 132 D with known galactic supernova remnants shows that they belong to this class and that they have resulted from type II supernovae (Westerlund and Mathewson, 1966). The three remnants are the results of explosions of young massive stars. It was estimated that the ages of the remnants are between 300 and 3000 yr; the uncertainty in this type of estimates is unavoidably great.

Westerlund and Mathewson also called attention to the fact that two of the remnants, N 49 and N 63 A, were found in a ring-like structure of neutral hydrogen and a series of stellar associations with a diameter of about 1000 parsecs. Inside the ring there is virtually no neutral hydrogen. They suggested that this shell is a result of an explosion of a super-supernova. The third supernova remnant, N 132 D, is situated on the inner edge of another ring of neutral hydrogen and stellar associations.

More recent radio data (presented at this Symposium by Mathewson) make it likely that the total number of supernova remnants in the Large Cloud should exceed 10. Their identification will contribute appreciably to answering the question "Could the associations and field stars in the Large Magellanic Cloud be formed as the result of a series of super-supernova explosions."

In the Small Magellanic Cloud (SMC) no optically identified supernova remnant exists so far. Large expanding shells of neutral hydrogen have been discussed by Hindman (1967). They may well be the results of gigantic explosions of similar type to what we have described as possibly having occurred in the Large Cloud. Of considerable interest is that recently Mathewson has found three non-thermal sources in the Small Cloud, and that all three are situated on the edges of the shells of neutral hydrogen.

We have previously (Westerlund, 1965) suggested that the Wing of the Small

Muller (ed.), The Magellanic Clouds, 31–33.

Magellanic Cloud may be the result of an explosion in the Small Cloud. From the Bar of the Small Cloud extends a narrow feature of extreme population I objects over a distance of about 7 kiloparsec. It is difficult to see how this feature could have been formed by tidal forces. Features similar to the Wing appear as 'jets' in many external galaxies.

Recently, Turner *et al.* (1968) have discussed the distribution of small clouds of neutral hydrogen between the two Magellanic Clouds. They propose that these H I clouds may have been ejected from the two galaxies.

2. Planetary Nebulae

Detailed discussions of the characteristics of the planetary nebulae in the two Clouds as well as of their distributions and velocities have recently been given by Westerlund (1968), Feast (1968) and Miss Webster (1969). We refer to those papers for more detailed information and summarize here only the most important results:

The distribution of the planetary nebulae in the SMC does not contradict the model of an edge-on galaxy. The radial velocities of the planetary nebulae do not indicate that the SMC is a rotating system.

The distribution of the planetary nebulae in the LMC is in agreement with a system being seen nearly face-on. The system is rotating; there is a disagreement about the centre of rotation.

A number of low-excitation nebulae in the SMC and a number of high-excitation nebulae in the LMC may have masses slightly higher than the average. The best value for the average hydrogen mass of a planetary nebula is near 0.1 solar mass.

In the LMC the planetary nebulae in the Western part are on the average less luminous than those in the Eastern part. At the same time, their nuclei contribute more to their total luminosity.

3. The Red Stars

Objective-prism surveys in infrared light of the Large Magellanic Cloud has led to the identification of about 600 M supergiants stars, about 400 carbon stars, and a number of possible population II red giant stars (Westerlund, 1960, 1964). It has been shown that the infrared absolute magnitudes of the M supergiants and carbon stars agree with those of the corresponding galactic objects. A study of the bar of the Large Cloud showed that a number of late-type M stars appear at $I = 13$, which, provided they are members of the Large Cloud, corresponds to the absolute luminosities of population II stars.

The M supergiant stars are concentrated to areas rich in blue young stars; the most marked difference in distribution is that the red supergiants avoid regions rich in emission nebulosity.

The carbon stars are mainly found in the outer areas of the Large Cloud. Their distribution indicates extensive arcs or arms.

The content of red stars in the Small Cloud is practically unknown. Infrared objective prism plates do not reveal the richness of red supergiants and carbon stars seen in the Large Cloud.

However, much remains to be done in this field. The survey will be repeated and extended at La Silla and photometry of the objects will be carried out as far as possible.

References

Feast, W. M.: 1968, in D. E. Osterbrock and C. R. O'Dell (eds.), 'Planetary Nebulae', *IAU Symp.* **34**, 34.
Henize, K. G.: 1956, *Astrophys. J. Suppl. Ser.* **2**, 315.
Hindman, J. V.: 1967, *Australian J. Phys.* **20**, 147.
Mathewson, D. S. and Healey, J. R.: 1964, in F. J. Kerr and A. W. Rodgers (eds.), 'The Galaxy and the Magellanic Clouds', *IAU/URSI Symp.* **20**, 283.
Turner, K. C., Varsavsky, C., and Tuve, M. A.: 1968, Carnegie Inst. Ann. Rep. for 1967, p. 290.
Webster, B. L.: 1969, *Monthly Notices Roy. Astron. Soc.* **143**, 79, 97, 113.
Westerlund, B. E.: 1960, *Uppsala Astron. Obs. Ann.* **4**, No. 7.
Westerlund, B. E.: 1964, in F. J. Kerr and A. W. Rodgers (eds.), 'The Galaxy and the Magellanic Clouds', *IAU/URSI Symp.* **20**, 239.
Westerlund, B. E.: 1965, *Mt. Stromlo Symp. on the Magellanic Clouds*, p. 40.
Westerlund, B. E.: 1968, in D. E. Osterbrock and C. R. O'Dell (eds.), 'Planetary Nebulae', *IAU Symp.* **34**, 23.
Westerlund, B. E. and Mathewson, D. S.: 1966, *Montly Notices Roy. Astron. Soc.* **131**, 371.

COMPARISON OF THE CEPHEID VARIABLES IN THE MAGELLANIC CLOUDS AND THE GALAXY

CECILIA PAYNE-GAPOSCHKIN
Smithsonian Astrophysical Observatory, Cambridge, Mass., U.S.A.

The difference between the Magellanic Clouds, evident from their sizes, structures and stellar content, is well illustrated by a comparison of their variable stars. The representation of the principal types is given in Table I.

TABLE I

	Large Cloud	Small Cloud
Cepheid variables	1111	1155
Type II cepheids	17	3
RR Lyrae stars (foreground)	28	31
Long-period and cyclic	59	24
Long-period (foreground)	8	1
Irregular	321	62
Eclipsing	79	34

In comparing the variable star populations we must take account of the sizes and distances of the Clouds. If we adopt as true moduli for the Large and Small Clouds the values 18.45 and 18.85 given by Sandage and Tammann (1968), we can compare the dimensions of the systems (Table II).

TABLE II

	Large Cloud	Small Cloud
Apparent visual diameter (Bok, 1966)	6°	2°.5
Diameter of area populated by variable stars (assumed circular)	8°.33	3°.50
Apparent area (square degrees)	54.5	9.6
Area (square kiloparsecs)	39	10

The numbers of variable stars per square degree and per square kiloparsec are then as given in Table III.

The distribution of variables in each Cloud is far from uniform. The most striking difference between the Clouds lies in the relative numbers of cepheids. Type II cepheids are about equally represented. Irregular variables are rather better represented in the Large Cloud, long-period and eclipsing stars in the Small.

Muller (ed.), The Magellanic Clouds, 34–46.

TABLE III

	Large Cloud		Small Cloud	
	per sq. deg.	per sq. kpc	per sq. deg.	per sq. kpc
Cepheid variables	20.4	28.5	121.6	115.5
Type II cepheids	0.3	0.4	0.3	0.3
Long-period variables	1.1	1.5	2.5	2.4
Irregular variables	5.9	8.2	6.5	6.2
Eclipsing stars	1.4	2.0	3.6	3.4
RR Lyrae stars (foreground)	0.5	–	3.3	–
Long-period var. (foreground)	0.2	–	0.1	–

1. Gross Properties of Cepheid Variables

The number of cepheids discussed in the present paper is smaller in the Large Cloud by 2½% than that known in the Small. Many cepheids certainly remain to be discovered in both, but probably the Large Cloud has been less completely searched. More than a hundred cepheids recently noted on Harvard plates, and not included in the present discussion, will swing the *numerical* balance in favor of the Large Cloud.

TABLE IV

Frequency of logarithm of period

log *P*	Large Cloud		Small Cloud		log *P*	Large Cloud		Small Cloud	
	Asymm.	Symm.	Asymm.	Symm.		Asymm.	W Vir	Asymm.	W Vir
−0.35 – −0.30	0	0	0	1	0.95 – 1.00	16	0	13	0
−0.30 – −0.25	0	0	0	1	1.00 – 1.05	19	0	15	0
−0.25 – −0.20	0	0	0	1	1.05 – 1.10	21	1	13	0
−0.15 – −0.10	0	0	0	2	1.10 – 1.15	23	0	14	0
−0.10 – −0.05	0	0	0	4	1.15 – 1.20	13	1	17	1
−0.05 – 0.00	0	0	0	2	1.20 – 1.25	15	1	11	0
0.00 – 0.05	0	0	1	5	1.25 – 1.30	6	0	4	0
0.05 – 0.10	0	0	17	5	1.30 – 1.45	11	1	3	0
0.10 – 0.15	0	2	38	10	1.35 – 1.40	10	0	3	0
0.15 – 0.20	1	1	57	11	1.40 – 1.35	8	0	4	0
0.20 – 0.25	3	4	73	11	1.45 – 1.50	7	3	6	1
0.25 – 0.30	2	7	64	29	1.50 – 1.55	3	1	7	0
0.30 – 0.35	4	4	55	13	1.55 – 1.60	5	4	2	0
0.35 – 0.40	28	15	57	12	1.60 – 1.65	1	2	2	1
0.40 – 0.45	55	9	65	6	1.65 – 1.70	4	1	1	0
0.45 – 0.50	107	11	98	15	1.70 – 1.75	1	1	0	0
0.50 – 0.55	134	12	75	6	1.75 – 1.80	0	0	0	0
0.55 – 0.60	114	3	66	1	1.80 – 1.85	0	0	2	0
0.60 – 0.65	109	11	45	0	1.85– 1.90	1	0	1	0
0.65 – 0.70	93	9	55	0	1.90 – 1.95	0	0	1	0
0.70 – 0.75	56	0	34	0	1.95 – 2.00	1	0	0	0
0.75 – 0.80	36	0	35	0	2.00 – 2.05	0	0	1	0
0.80 – 0.85	44	0	24	0	2.05 – 2.10	1	0	0	0
0.85 – 0.90	33	0	23	0	2.10 – 2.15	1	0	1	0
0.90 – 0.95	36	0	16	0	2.30 – 2.35	0	0	1	0

But the Small Cloud contains six times as many cepheids *per square degree*, and future discoveries are most unlikely to remove this difference.

The greatest excess of cepheids in the Small Cloud is contributed by those of short period. Stars of period greater than ten days are more numerous in the Large Cloud by a factor of 1.43, but the Small Cloud still has four times as many such cepheids per square degree, and nearly three times as many per square kiloparsec.

TABLE V

Frequency of $\langle m \rangle_0$ for cepheid variables

$\langle m \rangle_0$	Large Cloud	Small Cloud	$\langle m \rangle_0$	Large Cloud	Small Cloud
12.00–12.19	0	2	154.0–15.59	67	28
12.20–12.39	2	0	15.60–15.79	82	42
12.60–12.79	2	2	15.80–15.99	138	50
12.80–12.99	4	1	16.00–16.19	135	83
13.00–13.19	1	0	16.20–16.39	159	117
13.20–13.39	2	0	16.40–16.59	143	147
13.40–13.69	3	1	16.60–16.79	82	147
13.60–13.79	8	1	16.80–16.99	28	130
13.80–13.99	10	3	17.00–17.19	9	152
14.00–14.19	10	2	17.20–17.39	1	88
14.20–14.39	19	10	17.40–17.59	0	43
14.40–14.59	21	7	17.60–17.79	0	16
14.60–14.79	21	10	17.80–17.99	0	7
14.80–14.99	40	20	18.00–18.19	0	2
15.00–15.19	47	21	18.20–18.39	0	1
15.20–15.39	77	27	18.40–18.59	0	1

A significant comparison of the cepheids involves not only numbers, but gross and detailed properties and distribution. The gross properties comprise the frequency of period (or its logarithm) and of apparent magnitude. Detailed properties involve the amplitudes and forms of the light curves. Both gross and detailed properties differ in the two Clouds.

Table IV compares the frequency of $\log P$. As in the previous study of the Small Cloud by Payne-Gaposchkin and Gaposchkin (1966), stars with 'symmetrical' light curves (*M-m* over $0\overset{P}{.}3$) are tabulated separately. Table V gives the frequency of $\langle m \rangle_0$, the averaged magnitude at mean intensity, corrected for absorption. Both normal and symmetrical light curves are included. Parts of Tables IV and V are repeated from the study of the Small Cloud by Payne-Gaposchkin and Gaposchkin (1966).

The data of Tables IV and V furnish a numerical comparison of the gross properties. The period frequencies can be compared directly. The frequencies of $\langle m \rangle_0$, are affected by the difference of distance modulus for the two systems. We anticipate a later result, which obtains a value of $-0\overset{m}{.}27$ for the difference of distance moduli, LMC-SMC. The 'corrected' median value of $\langle m \rangle_0$ for the Small Cloud is the value that would be obtained if it were at the same distance as the Large Cloud (Table VI).

The difference in period frequency between the Clouds has long been known. Our median period for the Large Cloud is not much below the value $4\overset{d}{.}42$ given by Shapley and McKibben (1940) on the basis of 106 cepheids then known. No cepheids with periods less than $2\overset{d}{.}5$ had been found at that time; our list contains 67 such cepheids. When periods were available for 550 cepheids, the median value for the Large Cloud

TABLE VI

	Large Cloud			Small Cloud		
	Normal curves	Symmetrical curves	All	Normal curves	Symmetrical curves	All
Median log P (days)	0.630	0.459	0.614	0.493	0.266	0.470
Corresponding P (days)	4.266	2.877	4.111	3.112	1.845	2.915
Median $\langle m \rangle_0$, observed			16.00			16.61
corrected						16.34

was nearly the same, as may be seen from the data of Shapley and Nail (1955). As can be seen, our results from twice this number of cepheids have not resulted in a reduction of the median period.

The difference in period-frequency is certainly real. If the Large Cloud contained cepheids of short period that were similar in numbers and in light curves to those in the Small Cloud, there is no reason to think that they would have been missed. They should be brighter in apparent magnitude by $0\overset{m}{.}27$ on account of the difference in distance. The Harvard plate material for the two Clouds is strictly comparable, and is not inferior in quality or coverage for the Large Cloud. It was secured with the same instruments and has been studied by the same investigators, using the same methods. The Large Cloud is nearly 10° farther from the South Pole, and can be photographed at greater altitude and for a longer season than the Small, a factor that offsets inferior weather later in the season.

As will appear later, the amplitudes of the cepheids of shortest period in both Clouds tend to be small, and the question has been raised whether this has militated against the discovery of cepheids of short period. We therefore make an approximate correction for this effect.

Kukarkin (1954) has applied estimated corrections for discovery chance to the observed frequency of log P. His adopted discovery probabilities are functions of amplitude, and range from 0.73 to 0.95 for amplitudes $0\overset{m}{.}76$ to $1\overset{m}{.}31$. Table VII shows the results of applying these corrections to the frequencies of Table IV, and is given in intervals of 0.1 in log P and in percentages for comparison with Kukarkin. Results are tabulated separately for asymmetric curves and for all cepheids. To obtain the latter values, the numbers of symmetrical curves (of small amplitude) were corrected separately for discovery chance, and Kukarkin's factors were arbitrarily extrapolated to include discovery chance 0.55 for amplitude $0\overset{m}{.}61$. The comparison is summarized in Table VIII.

TABLE VII

Frequency of Period, Corrected for discovery chance

Mean log *P*	Large Magellanic Cloud				Small Magellanic Cloud					
	Observed		Corrected		Observed		Corrected		Kukarkin
	Asymm. %	All. %	Asymm. %	All %	Asymm. %	All %	Asymm. %	All %	Corrected %
0.05	0.0	0.0	0.0	0.0	1.8	2.4	2.1	3.0	2.3
0.15	0.1	0.4	0.2	0.7	9.3	10.1	10.1	11.2	6.3
0.25	0.5	1.4	0.6	2.4	13.4	15.5	13.9	16.7	9.6
0.35	3.1	4.6	3.2	5.9	11.0	12.0	11.0	12.4	13.4
0.45	15.9	16.4	16.4	16.3	16.0	16.1	15.5	15.4	13.4
0.55	14.5	14.7	25.0	23.8	13.8	12.9	13.6	12.4	12.4
0.65	19.8	20.0	19.9	20.3	9.8	8.6	9.5	8.0	8.9
0.75	8.9	8.3	8.8	7.5	6.8	5.9	6.6	5.6	7.3
0.85	7.5	6.9	7.3	6.2	4.6	4.0	4.7	4.0	6.3
0.95	5.1	4.7	5.0	4.3	2.8	2.4	2.9	2.5	5.3
1.05	3.9	3.6	3.8	3.2	2.7	2.4	2.7	2.3	4.3
1.15	3.5	3.2	3.2	2.7	3.0	2.6	2.9	2.5	3.5
1.25	2.1	1.9	1.8	1.6	1.5	1.3	1.3	1.1	2.8
1.35	2.1	1.9	1.8	1.6	0.6	0.5	0.5	0.4	1.8
1.45	1.5	1.4	1.2	1.1	1.0	0.9	0.9	0.7	1.0
1.55	0.8	0.7	0.7	0.6	0.9	0.8	0.8	0.7	0.8
1.65	0.5	0.4	0.4	0.4	0.3	0.3	0.3	0.2	0.5
1.75	0.1	0.1	0.1	0.1	–	–	–	–	–
1.85	0.1	0.1	0.1	0.1	0.3	0.3	0.3	0.2	–
1.95	0.1	0.1	0.1	0.1	0.1	0.1	0.1	0.1	–
2.05	0.1	0.1	0.1	0.1	0.1	0.1	0.1	0.1	–
2.15	0.1	0.1	0.1	0.1	0.1	0.1	0.1	0.1	–
2.35	–	–	–	–	0.1	0.1	0.1	0.1	–

The values of median log *P* have been slightly reduced by applying the correction for discovery chance. However, the *difference* in median log *P* has not been reduced. The difference between the results of Shapley and of Kukarkin for the Small Cloud and the values deduced from our data reflects the increase in the number of known short periods in the Small Cloud; by doubling the number of stars we have *increased*

TABLE VIII

	Large Cloud		Small Cloud	
	Median log *P*	Corr. *P*	Median log *P*	Corr. *P*
Asymmetric curves, as observed	0.630	4.266	0.493	3.112
Asymmetric curves, corrected	0.625	4.217	0.480	3.020
All curves, corrected	0.600	4.027	0.422	2.642
Shapley (1942)			0.537	3.443
Corrected and smoothed (Kukarkin)			0.540	3.467

the difference in period frequency between the Clouds. We note that the correction for discovery chance is uncertain, depending as it does on the method of discovery and the number of comparisons on which it is based. Probably the discovery chance for stars of small amplitude is greater than Kukarkin estimated, and his factors would then lead to overcorrection.

Wesselink (1966) has called attention to two selection effects that may operate on the observed period frequency: discovery chance decreases toward the plate limit, and is reduced by increasing background density of faint stars. The plate limit effect would *favor* the discovery of (absolutely) faint variable stars in the Large Cloud on account of the difference of distance modulus, and should thus favor the discovery of cepheids of short period. The effect of background density should be greatest in the crowded 'bar' regions of the Large Cloud. Median $\log P$ for 339 cepheids in the bar is 0.632, as compared to 0.630 for all asymmetric curves in the Large Cloud. These numbers do not suggest that the number of discoveries of stars of short period has been affected by high star density in the bar.

We therefore consider it established that the difference of period frequency between the Clouds is real. Differences of period frequency in different regions of the Clouds are also real and significant.

2. Detailed Properties of Cepheid Variables

Data for comparison of the detailed properties are contained in Tables IX and X. Table IX relates to asymmetric light curves. Successive columns contain: (1) mean logarithm of period; (2) magnitude at maximum, M_0; (3) magnitude at minimum, m_0; (4) magnitude at mean intensity, $\langle m \rangle_0$; (5) amplitude A in magnitudes; (6) interval R from increasing mean magnitude to maximum; (7) interval W from increasing to decreasing mean magnitude; (8) luminosity moment S, defined by Christy (1968, p. 30) as 'the delay after mean light of the center of the positive portion of the luminosity curve'; (9) interval M_1-m from minimum to the following maximum; (10) interval M_2-m when two maxima are present; and (11) M_2-M_1, interval between maxima. An asterisk denotes the brighter maximum. Data for periods up to 10 days are combined in groups of 50 stars, for periods from 10 to 30 days, in groups of 20, for periods over 30 days, in groups of 10 or less. The first entry comprises 10 stars for the Large Cloud, 5 for the Small. See Figure 1.

Table X, for stars with symmetrical light curves, is similar to Table IX except for the omission of columns (10) and (11), since these stars do not show secondary maxima. The data are combined in groups of 20 stars or less.

All tabulated magnitudes are corrected for absorption. The quantities R, W, S, M-m and M_2-M_1 are expressed in terms of the period: R, M, and S can be determined more objectively than M-m, since the exact phase of minimum is often hard to define, and falls in the least well observed part of the light curve.

Tables IX and X show that there are systematic differences, intrinsic differences, and similarities between the cepheids in the two Clouds.

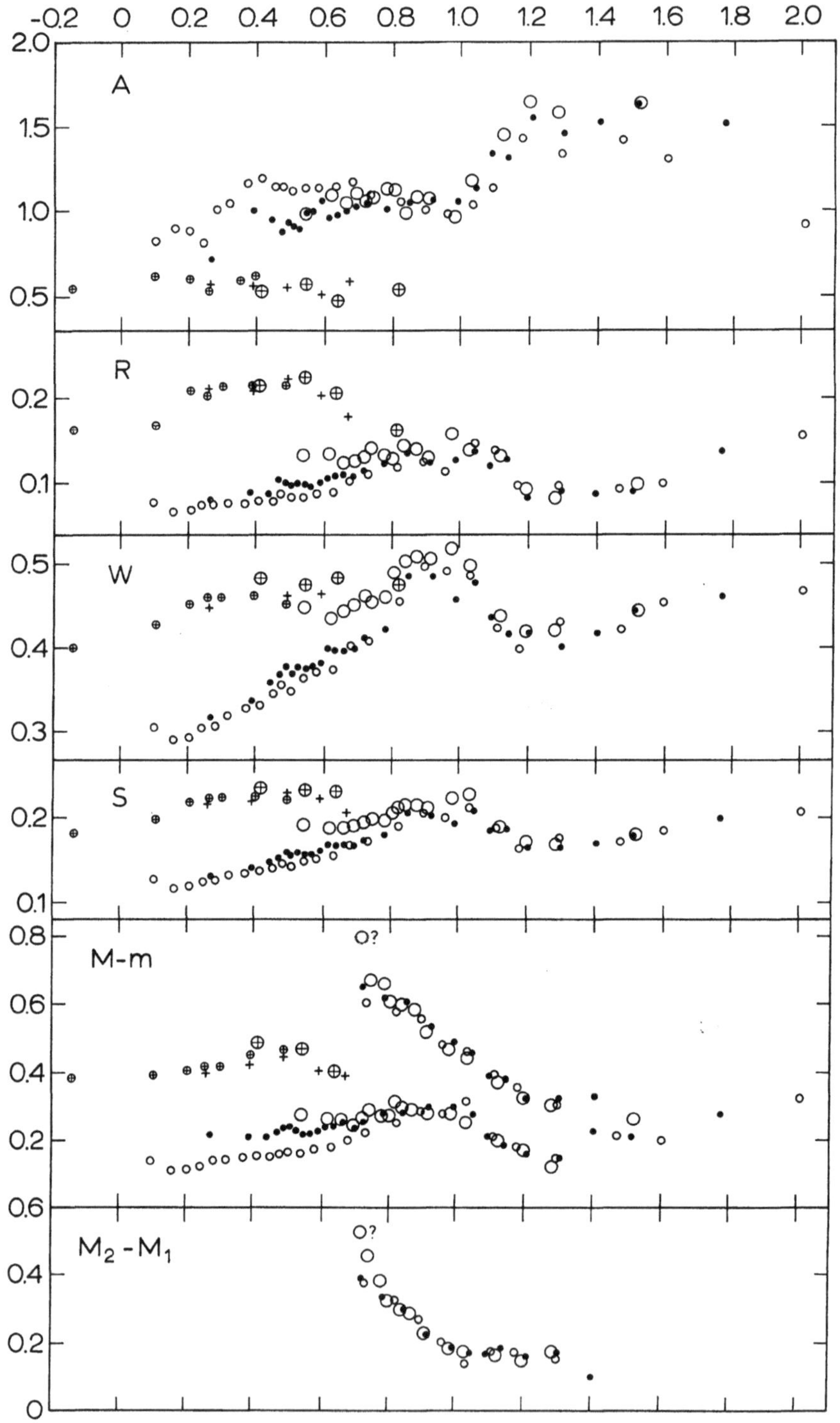

Fig. 1.

TABLE IX

Cepheid light curve parameters

Large Cloud

log P	M_0	m_0	$\langle m\rangle_0$	A	R	W	S	M_1-m	M_2-m	M_2-M_1
0.2708	16.48	17.22	16.88	0.73	0.082	0.317	0.133	0.220	–	–
0.3952	15.97	16.99	16.55	1.02	0.090	0.339	0.143	0.216	–	–
0.4444	15.91	16.87	16.46	0.96	0.089	0.360	0.150	0.212	–	–
0.4713	15.92	16.81	16.42	0.89	0.105	0.369	0.155	0.229	–	–
0.4920	15.87	16.81	16.41	0.94	0.101	0.379	0.160	0.240	–	–
0.5098	15.85	16.77	16.38	0.92	0.100	0.371	0.157	0.243	–	–
0.5283	15.83	16.73	16.34	0.90	0.102	0.379	0.160	0.232	–	–
0.5492	15.70	16.71	16.28	1.01	0.100	0.376	0.159	0.224	–	–
0.5692	15.64	16.65	16.21	1.01	0.098	0.379	0.159	0.220	–	–
0.5937	15.52	16.59	16.12	1.07	0.102	0.384	0.162	0.232	–	–
0.6167	15.39	16.36	15.94	0.97	0.108	0.401	0.170	0.244	–	–
0.6405	15.49	16.48	16.02	0.99	0.110	0.399	0.170	0.243	–	–
0.6669	15.43	16.44	15.98	1.01	0.112	0.398	0.170	0.257	–	–
0.6911	15.33	16.37	15.88	1.04	0.110	0.400	0.170	0.247	–	–
0.7289	15.21	16.27	15.79	1.06	0.117	0.414	0.177	0.253*	0.651	0.398
0.7856	14.95	15.98	15.50	1.03	0.125	0.424	0.183	0.282*	0.622	0.340
0.8509	14.86	15.93	15.39	1.07	0.139	0.486	0.208	0.297	0.602	0.305
0.9216	14.73	15.82	15.25	1.09	0.127	0.487	0.205	0.305*	0.538	0.233
0.9952	14.53	15.60	15.03	1.07	0.130	0.459	0.196	0.301	0.494	0.193
1.0500	14.38	15.54	14.94	1.16	0.152	0.480	0.211	0.284	0.462*	*0.178
1.0962	14.26	15.62	15.01	1.36	0.123	0.438	0.187	0.218	0.396*	0.178
1.1416	14.15	15.48	14.96	1.33	0.130	0.418	0.189	0.190	0.378*	0.197
1.2099	13.86	15.43	14.72	1.57	0.086	0.419	0.168	0.164	0.330*	0.166
1.3030	13.86	15.43	14.72	1.48	0.093	0.404	0.166	0.154	0.332*	0.178
1.4089	13.31	14.86	14.25	1.55	0.092	0.420	0.178	0.231	0.338*	0.107
1.5165	12.73	14.38	13.60	1.65	0.093	0.447	0.130	0.217	–	–
1.7749	12.46	13.88	13.11	1.44	0.143	0.464	0.202	0.285	–	–

Small Cloud

log$P$	M_0	m_0	$\langle m\rangle_0$	A	R	W	S	M_1-m	M_2-m	M_2-M_2
—0.1208	17.11	17.76	17.45	0.65	0.143	0.350	0.164	0.352	–	–
0.1029	16.84	17.67	17.37	0.83	0.078	0.306	0.128	0.146	–	–
0.1637	16.66	17.56	17.28	0.90	0.068	0.292	0.119	0.118	–	–
0.2094	16.57	17.46	17.15	0.89	0.070	0.293	0.121	0.120	–	–
0.2464	16.59	17.41	17.15	0.82	0.075	0.306	0.127	0.128	–	–
0.2876	16.41	17.43	17.05	1.02	0.076	0.308	0.128	0.144	–	–
0.3249	16.36	17.42	17.06	1.06	0.078	0.321	0.133	0.149	–	–
0.3783	16.10	17.27	16.85	1.17	0.078	0.329	0.136	0.153	–	–
0.4194	15.96	17.16	16.75	1.20	0.081	0.334	0.138	0.157	–	–
0.4551	15.99	17.15	16.76	1.16	0.080	0.348	0.143	0.155	–	–
0.4799	15.91	17.07	16.64	1.16	0.088	0.357	0.148	0.166	–	–
0.5081	15.93	17.06	16.68	1.13	0.085	0.350	0.145	0.169	–	–
0.5440	15.77	16.92	16.52	1.15	0.085	0.367	0.151	0.164	–	–
0.5826	15.71	16.86	16.42	1.15	0.090	0.374	0.155	0.178	–	–
0.6324	15.64	16.80	16.37	1.16	0.091	0.377	0.156	0.181	–	–
0.6839	15.49	16.68	16.20	1.19	0.105	0.405	0.170	0.204	–	–
0.7394	15.50	16.61	16.13	1.11	0.114	0.411	0.175	0.227*	0.610	0.383
0.8237	15.35	16.42	15.92	1.07	0.121	0.457	0.193	0.251*	0.585	0.334
0.9000	15.14	16.16	15.62	1.02	0.128	0.500	0.209	0.290*	0.566	0.276
0.9614	15.11	16.11	15.60	1.00	0.118	0.492	0.203	0.283*	0.488	0.205
1.0368	14.69	15.74	15.22	1.05	0.150	0.488	0.213	0.320	0.466*	0.146
1.1096	14.56	15.71	15.16	1.15	0.141	0.430	0.190	0.216	0.398*	0.180
1.1796	14.40	15.85	15.20	1.45	0.100	0.401	0.167	0.184	0.364*	0.180
1.2996	14.13	15.49	14.84	1.36	0.101	0.434	0.178	0.151	0.314*	0.163
1.4788	13.75	15.19	14.52	1.44	0.098	0.426	0.175	0.222	–	–
1.6010	13.46	14.79	14.12	1.33	0.104	0.457	0.187	0.207	–	–
2.0161	12.21	13.16	12.64	0.95	0.162	0.470	0.211	0.330	–	–

TABLE X

Cepheid light curve parameters, symmetrical curves

Large Cloud								
log *P*	M_0	m_0	$\langle m \rangle_0$	*A*	*R*	*W*	*S*	*M-m*
0.2652	15.96	16.53	16.22	0.57	0.210	0.449	0.220	0.407
0.3918	15.82	16.39	16.06	0.57	0.212	0.468	0.227	0.422
0.4914	15.40	15.96	15.68	0.56	0.223	0.463	0.229	0.456
0.5946	15.47	16.00	15.80	0.53	0.204	0.465	0.223	0.404
0.6749	15.45	16.05	15.57	0.60	0.180	0.448	0.209	0.394
Small Cloud								
log *P*	M_0	m_0	$\langle m \rangle_0$	*A*	*R*	*W*	*S*	*M-m*
0.1357	16.99	17.54	17.24	0.55	0.164	0.401	0.183	0.382
0.1098	16.56	17.19	16.88	0.63	0.169	0.431	0.200	0.394
0.2104	16.48	17.08	16.78	0.60	0.211	0.453	0.221	0.410
0.2646	16.27	16.81	16.56	0.54	0.205	0.461	0.222	0.420
0.3034	16.30	16.90	16.60	0.60	0.216	0.461	0.226	0.421
0.4004	16.18	16.82	16.48	0.64	0.219	0.464	0.228	0.456
0.4950	15.70	16.32	16.01	0.62	0.217	0.454	0.224	0.464

At similar periods the apparent magnitudes differ *systematically*; this difference we ascribe to a difference of distance modulus. A graphical evaluation of the difference, for the range of periods common to the two Clouds, indicates that the modulus of the Large Cloud is smaller by $0^m.27$, if we assume that the Clouds have a common period-luminosity relation. This difference depends critically on the standard magnitudes used: those of Arp (1960) for the Small Cloud, and those of Hodge (1961) for the Large. It depends to an even greater extent on the adopted corrections for absorption, described in previous papers by Payne-Gaposchkin and Gaposchkin (1966) and Payne-Gaposchkin (1968). As noted above, Sandage and Tammann (1967) adopt a systematic difference of $0^m.40$ in the same sense.

When the systematic differences are eliminated, we find a number of *intrinsic* differences:

1. Average amplitude, for stars of period under 10 days, is greater in the Small Cloud; for periods over 10 days, it is greater in the Large Cloud.

2. The interval *R* between mean magnitude and maximum brightness is greater in the Large Cloud for periods under 10 days, smaller for periods over 10 days. The rate of brightening *R* and the amplitude *A* are in fact correlated: curves of greater *A* have smaller *R* and are therefore steeper. The interval *M-m*, from minimum to maximum, though less accurately determined than *R*, shows the same contrast in the two Clouds.

3. The width *W* of the light curve at mean magnitude is greater in the Large Cloud for periods less than about 5 days; for longer periods the values of *W* are sensibly the same.

We note that *R* and *W* are measured with reference to mean magnitude, not mean intensity. If they had been referred to mean intensity the observed differences between the Clouds would in every case have been increased (since both R and W would then become smaller for larger amplitudes), on account of the differences in amplitude at corresponding periods in the two Clouds. The change would in no case exceed 10%, and would usually be much smaller.

Values of the luminosity moment S are slightly larger in the Large Cloud for periods up to about 5 days; for longer periods the values do not differ sensibly in the two Clouds.

4. The period-luminosity relations at maximum, minimum and mean intensity are shown in Figure 2. The lowest section of Figure 2 combines the data for the two Clouds by subtracting the difference, $0^{m}.27$, from the magnitudes for the Small Cloud. For values of log P less than 1.2 there is little evidence of any difference; for longer periods the slope may be somewhat larger for the Large Cloud, but the scatter is great and the number of stars comparatively small. As was already pointed out for the Small Cloud, the data as they stand suggest that the average relation is not linear, but this effect may be produced by incompleteness of the material for the faintest cepheids in the Small Cloud; the corrections applied for absorption also introduce great uncertainties. In view of the accidental and intrinsic dispersion of the period-luminosity relation, I believe that we are justified in supposing that the Clouds can be regarded as having a common period-luminosity relation. This belief is also expressed by Sandage and Tammann (1967), although the form of the relation that they favor is not necessarily the same as the one indicated by our data. The relation between $\langle m \rangle_0$ and log P for all stars in both Clouds (SMC adjusted by $0^{m}.27$) is shown in bottom part of Figure 2.

There is one definite *similarity* between the light curves in the two Clouds. The interval M_2-M_1 is the same at a given period, and the occurrence of two maxima covers the same interval of period in both systems. The similarity of width W and of luminosity moment S for periods over 5 days is probably caused by the same phenomenon that produces the secondary maximum, even at periods shorter than those at which the second maximum can be observed as a separate entity.

The properties of the symmetrical curves do not differ sensibly in the two Clouds (apart from the systematic difference of magnitude attributed to a difference of distance), though their period frequency is not the same.

The light curves of galactic cepheids are described in Table XI. The parameters are derived from the data for 190 photoelectrically observed galactic cepheids, principally derived from a discussion of the tabulation of Mitchell, Iriarte, Steinmetz and Johnson (1964). The amplitudes are on the B system, with which our own system based on the Bruce plates is probably not identical, so that the comparison of amplitudes is not precise. It can be concluded, however, that the amplitudes of galactic cepheids at a given period match those of the Large Cloud more closely than those of the Small over the whole range of periods common to these systems. The interval M_2-M_1 has the same relation to period as in the Clouds. The parameters for Galactic cepheids are included in Figure 1.

Table XII summarizes the comparison of light curves for galactic and Magellanic cepheids. Asymmetric light curves only are considered. No difference is apparent in the parameters for symmetrical light curves.

All the differences for periods under 10 days are aspects of the systematic differences

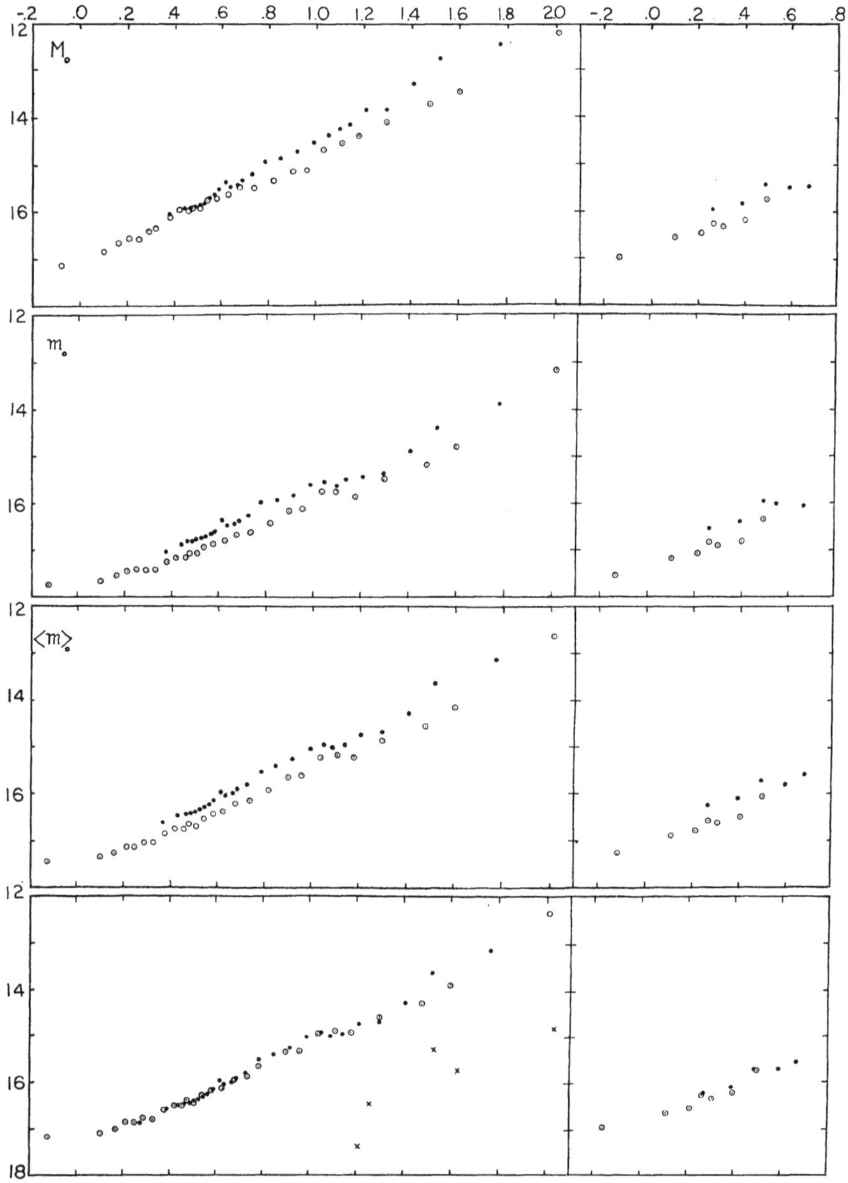

Fig. 2. Logarithm of period and magnitude at maximum, minimum, and mean intensity. Dots, Large Cloud; circles, Small Cloud. Bottom: magnitude at mean intensity, with magnitudes for Small Cloud corrected by -0.27. W Virginis stars, crosses. Right hand section, symmetrical light curves.

TABLE XI

Mean light curve parameters for galactic cepheids

Asymmetric light curves

log *P*	*A*	*R*	*W*	*S*	M_1-*m*	M_2-*m*	M_2-M_1
0.5419	1.00	0.136	0.450	0.195	0.281	–	–
0.6208	1.11	0.137	0.437	0.191	0.268	–	–
0.6635	1.06	0.127	0.447	0.191	0.268	–	–
0.6960	1.12	0.129	0.453	0.194	0.248	–	–
0.7209	1.07	0.132	0.463	0.198	0.272*	0.802:	0.530:
0.7469	1.10	0.144	0.458	0.201	0.296*	0.674	0.468
0.7816	1.14	0.136	0.461	0.199	0.280*	0.670	0.390
0.8074	1.14	0.132	0.492	0.208	0.280*	0.612	0.332
0.8400	1.00	0.147	0.505	0.217	0.300*	0.608	0.308
0.8734	1.10	0.143	0.509	0.217	0.298*	0.590	0.292
0.9016	1.09	0.132	0.507	0.213	0.287*	0.524	0.237
0.9815	0.98	0.162	0.520	0.227	0.284	0.474*	0.190
1.0308	1.20	0.142	0.500	0.231	0.260	0.445*	0.185
1.1217	1.47	0.136	0.440	0.192	0.203	0.377*	0.174
1.2002	1.66	0.098	0.422	0.173	0.180	0.337*	0.157
1.2865	1.60	0.087	0.423	0.170	0.130	0.312*	0.182
1.5220	1.66	0.104	0.447	0.184	0.270	–	–

Symmetrical light curves

log *P*	*A*	*R*	*W*	*S*	*M*-*m*
0.4184	0.54	0.129	0.485	0.235	0.490
0.5452	0.58	0.226	0.477	0.234	0.471
0.6426	0.49	0.209	0.485	0.231	0.406
0.8215	0.55	0.166	0.476	0.214	0.315

of amplitude: large amplitudes go with rapid rise of brightness, small *M*-*m* and large *S*.

The systematic difference of amplitude is clearly related to the difference in period frequency. Large amplitudes are attained at log $P=0.4$ for the Small Cloud, about 0.6 for the Large Cloud, and about 0.8 for the Galaxy. We recall that median log *P* for the Small Cloud is 0.422, for the Large Cloud, 0.625. The corresponding figure for

TABLE XII

Comparison of light curve parameters for Magellanic and Galactic cepheids

Parameter	Period under 10 days			Period over 10 days		
	Galaxy	Large Cloud	Small Cloud	Galaxy	Large Cloud	Small Cloud
R	largest	intermediate	smallest		equal	
W	largest	intermediate	smallest		equal	
S	largest	intermediate	smallest		equal	
A	intermediate:	smallest	largest	largest	intermediate	smallest
M-*m*	largest	intermediate	smallest		equal	
M_2-M_1		equal			equal	

the Galaxy is less certain. A rough estimate for the photoelectrically observed cepheids (complete to about the twelfth apparent magnitude), obtained by applying extinction corrections and correcting for volume coverage within a flattened circular area, is 0.785. In any case it is generally conceded that the median period for the galactic cepheids of our neighborhood is greater than for either Cloud.

Table XIII suggests that the interpretation of period frequency and of amplitude are aspects of a single phenomenon.

TABLE XIII

	Galaxy	Large Cloud	Small Cloud
Median log *P*	0.785:	0.635	0.422
Log *P* for maximum *A*	0.8	0.6	0.4

References

Arp, H. C.: 1960, *Astron. J.* **65**, 404.
Bok, B. J.: 1966, *An. Rev. Astron. Astrophys.* **4**, 95.
Christy, R. W.: 1968, *Q. J. Roy. Astron. Soc.* **9**, 13.
Hodge, P. W.: 1961, *Astrophys. J. Suppl.* **6**, 235.
Kukarkin, B. W.: 1954, *Erforschung der Struktur und Entwicklung der Sternsysteme auf der Grundlage des Studiums Veränderlicher Sterne*, p. 38.
Mitchell, R. I., Iriarte, B., Steinmetz, D., and Johnson, H. L.: 1964, *Bol. de las Obs. Ton. y Tac* **3**, 153.
Payne-Gaposchkin, C.: 1968, *Smithsonian Contrib. Astrophys.*, in press.
Payne-Gaposchkin, C. and Gaposchkin, S.: 1966, *Smithsonian Contrib. Astrophys* **9**.
Sandage, A. and Tammann, G. A.: 1968, *Astrophys. J.* **151**, 531.
Shapley, H. and McKibben, V.: 1940, P.N.A.S. **26**, 105.
Shapley, H. and Nail, V. McK.: 1955, P.N.A.S. **41**, 829.
Wesselink, A. J.: 1966, *Astron. J.* **71**, 185.

THE SHORT-PERIOD VARIABLE STARS IN THE MAGELLANIC CLOUDS

JORGE LANDI DESSY
Observatorio Astronómico e IMAF, Universidad de Córdoba

and

JOSÉ R. LABORDE
Observatorio Astronómico, Universidad de Córdoba, Argentina

About twenty years ago the problem arose that in the Clouds no variables were known with a period shorter than one day, therefore Córdoba Observatory began a search for stars fainter than had been studied up to then at Harvard Observatory. Blinking 28 pairs of plates an enormous number of variables was found and among them a significative number with periods between $0\overset{d}{.}2$ and $1\overset{d}{.}5$, whose magnitudes were between 17th and 18th. At the meeting of the Galaxies Commission in Rome in 1952 I

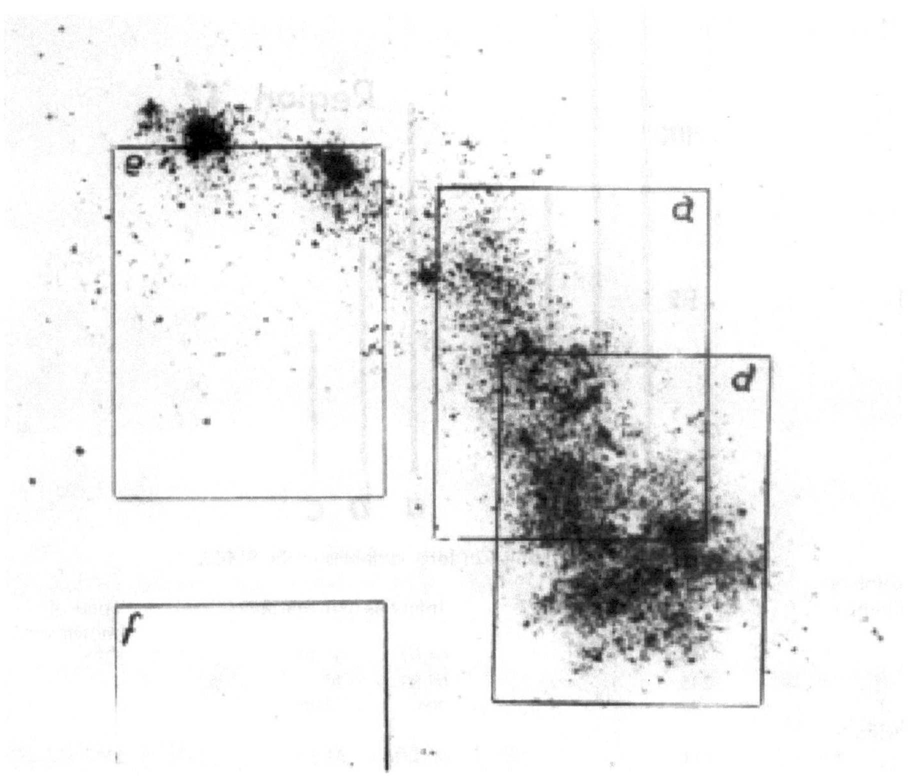

Muller (ed.), The Magellanic Clouds, 47–49.

announced this discovery and at the same time Thackeray reported to have found variables of the RR Lyrae type but with a magnitude of 19. Since then the location of both groups and their belonging to the Clouds has been discussed.

After Wesselink's work who confirmed the existence of the brightest short-period variables in the Small Cloud – we arrived at the conclusion that they form a special group connected rather with population I, so that they were not the expected RR

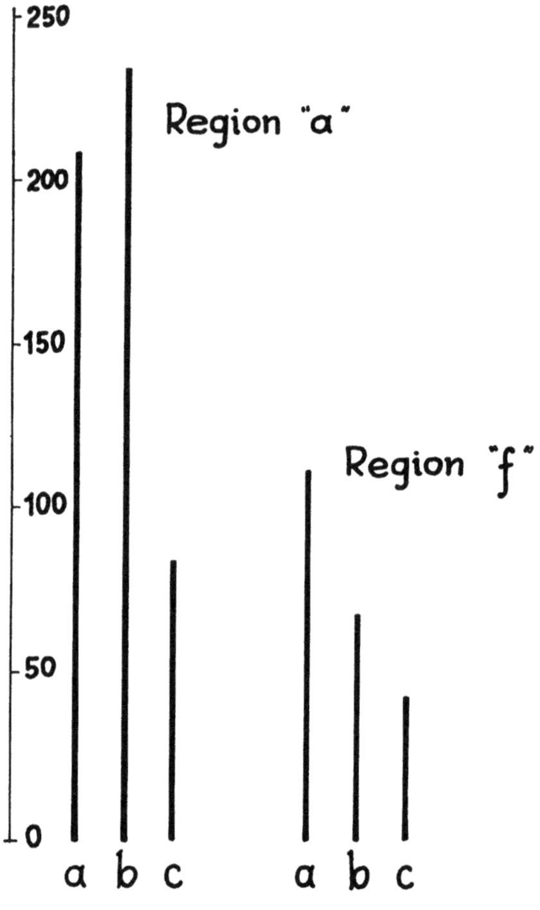

Fig. 2. Harvard and Cordoba variables in the SMC.

Region 'a' Column	Number of variables	Intervals between plates		Total of known var.
a	209	$0^d.07$	– $0^d.29$	528
b	235	$0^d.83$	– 2^d	
c	84	20^d	– 1829^d	
Region 'f'				
a	111	$0^d.146$	– $0^d.27$	222
b	68	$0.^d95$	– 2^d	
c	43	2^d	– 3367^d	

Lyrae. Later studies of the Larger Cloud did not show the existence of this group of variables, at least not in the studied zone.

In order to find out whether these variables were present in other areas of the Small Cloud, plates were taken of a new region (Region 'f', Figure 1) quite far away from the bar, in hope to find variables and see if their distribution in type were the same as in the region 'a'. To our surprise, 222 stars were found of which 66 already had been discovered at Harvard. Distributing adequately the intervals between the plates for the blink, we found that 111 variables were on plates whose intervals were $0\overset{d}{.}27$ or less; 68 in pairs between $0\overset{d}{.}95$ and 2^d; and 43 stars in pairs with intervals longer than two days.

Of the 111 variables of the first group, 67 have also been found in pairs with intervals of less than $0\overset{d}{.}146$. If we observe the histograms of the region 'a' (Figure 2) of the Small Cloud, it seems that in the region 'f' are located a larger quantity of variables of rapid variation or rather, that bright long-period cepheids are missing there and, on the contrary, there exists a large number of variables with short periods and therefore with less brightness.

Unfortunately the climate at Bosque Alegre is not favourable for the study of variables of such short periods, but it is possible that some of the new instruments with aplanatic or anastigmatic optics which are being installed in very good seeing regions can be used for this problem even if only a few nights.

References

Bock, B.: 1966, 'Magellanic Clouds', *Ann. Rev. Astron. Astrophys.* **4**.

Landi Dessy, J.: 1959, *Publ. Astron. Soc.* **71**, 435.

Landi Dessy, J.: 1964, in F. J. Kerr and A. W. Rodgers (eds.), 'The Galaxy and the Magellanic Clouds', *IAU Symp.* **20**, 377.

Thackeray, A. D.: 1964, in F. J. Kerr and A. W. Rodgers (eds.), 'The Galaxy and the Magellanic Clouds', *IAU Symp.* **20**, 370.

Thackeray, A. D. and Wesselink, A. J.: 1953, *Nature* **171**, 693.

Van den Bergh, S.: 1968, 'The Galaxies of the Local Group', *R.A.S.C.J.* **62**, No. 4.

Wesselink, A. J.: 'Veränderlichen-Colloquium Bamberg', *Kleine Veröffentl. Remeis-Sternwarte Bamberg* **34**, 7–8.

PROPERTIES OF THE NEUTRAL HYDROGEN IN THE MAGELLANIC CLOUDS

F. J. KERR
Astronomy Program, University of Maryland, College Park, Md., U.S.A.

1. History

The first low-resolution survey of the neutral hydrogen in the Clouds (Kerr *et al.*, 1954) showed the existence of large H I envelopes around both Clouds and an unexpectedly large gas content for the Small Cloud. The second survey (Hindman *et al.*, 1963) was also carried out with a small telescope but a better receiver was used; this study confirmed the general features of the distribution and demonstrated that a bridge of gas extended right across between the Clouds (Figure 1).

The main detailed work has been done with the Australian 210-ft telescope at Parkes, for the LMC by McGee and Milton (1966) and for the SMC by Hindman (1967). The beamwidth of this telescope is 13 min arc, corresponding to a linear resolution of 230 pc at the distance of the Clouds. Turner has made recent observations in Argentina, including a detailed study of the bridge and some new data on the SMC.

2. General Appearance

The LMC hydrogen shows a very patchy appearance, while the SMC gas is much smoother in its distribution. The LMC has 52 clearly-defined gas concentrations, the SMC only three, but the two Clouds contain roughly the same total amount of gas. Figures 2 and 3 show the integrated surface brightness over all velocities; in order to trace out the concentrations properly, the velocity structure is required also. An average concentration in the LMC has a diameter of about 500 pc, a velocity dispersion of 11.5 km/sec, a gas density of 1 atom/cm^3 and a total mass of 4×10^6 $M_\odot$.

The LMC profiles sometimes show a single peak, but generally there are two or (in 10% of the cases) three peaks. Evidently we are seeing structure in depth, with gas concentrations behind one another (Figure 4). The SMC has double-peaked profiles over much of its area, again implying structure in depth (Figure 5). The bridge is continuous on the sky, but its detailed study indicates that it also is patchy in form and seems to be showing three-dimensional structure.

One of the surprises in the Magellanic Cloud studies has been the observation of brightness temperatures as high as 150 K or so at the peaks of the profiles in some places, in spite of the small sizes of these galaxies. These temperatures are almost as high as the highest observed anywhere in our Galaxy. We can conclude that the actual spin temperature is at least 150 K, and also that there are high column densities of hydrogen atoms in some areas. Unfortunately there are no strong continuum sources

Muller (ed.), The Magellanic Clouds, 50–65. All Rights Reserved.

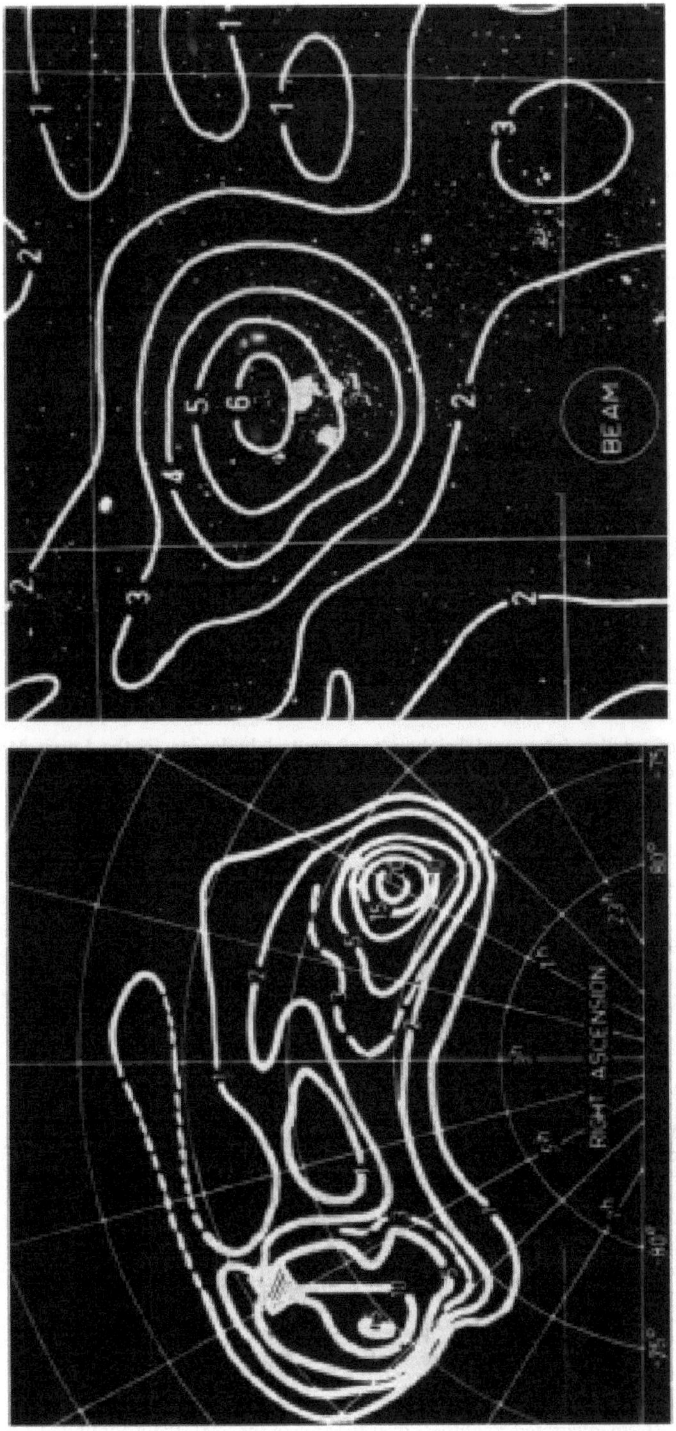

Fig. 1. Left: Contours of integrated brightness of neutral hydrogen in the Magellanic System from a low-resolution survey. Contour unit $= 2 \times 10^{-16}$ Wm^{-2} sr^{-1} (Hindman *et al.*, 1963). Right: Example of the increased detail provided by the 210-ft telescope, for the small region of the Large Cloud shown hatched in the other diagram (McGee, 1964). The radio contours are superposed on a star map of the Clouds.

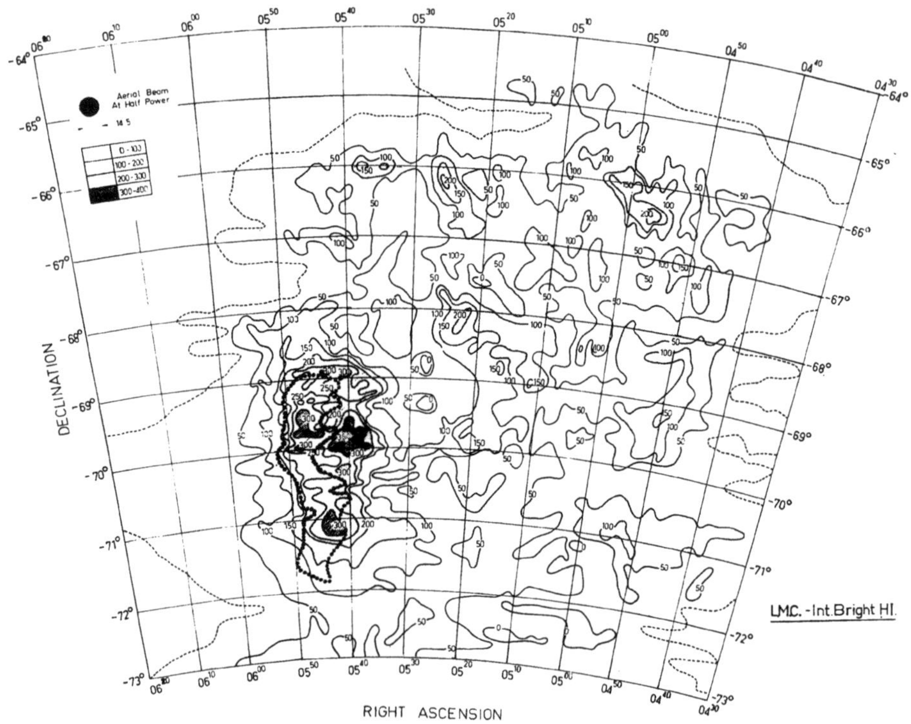

Fig. 2. Contours of relative integrated brightness of neutral hydrogen in the Large Magellanic Cloud. (McGee, R. X. and Milton, J. A.: 1964, *IAU Symp.* **20**, 291.)

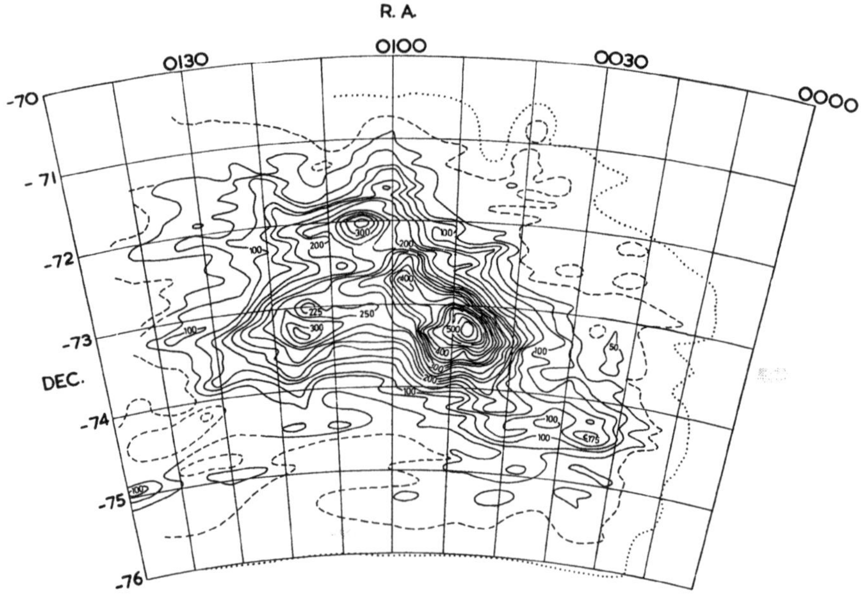

Fig. 3. Contours of 21-cm integrated brightness over the area of the SMC (contour unit $= 3.5 \times 10^{-17}$ $Wm^{-2}\ sr^{-1}$). The outer (dotted) contour is an estimate of the limits of detectable radiation with the present receiver sensitivity. Epoch of coordinates 1975 (Hindman, 1967).

(radio galaxies) behind the Clouds. If there were, observations of 21-cm absorption would provide useful information on the physical properties of the gas in the Clouds.

3. Relationships with Other Constituents

Close correlation is observed with H II and with supergiant stars, including the so-called constellations of Shapley (see Figure 6). In the LMC, McGee and Milton (1966) found positional agreement with H I complexes for 70 out of 90 Henize H II regions. Furthermore, in a study of objects with velocities measured at the Radcliffe Observatory, 54 out of 70 OB stars and all 42 H II regions have motions closely related to those of the corresponding H I (McGee, 1964).

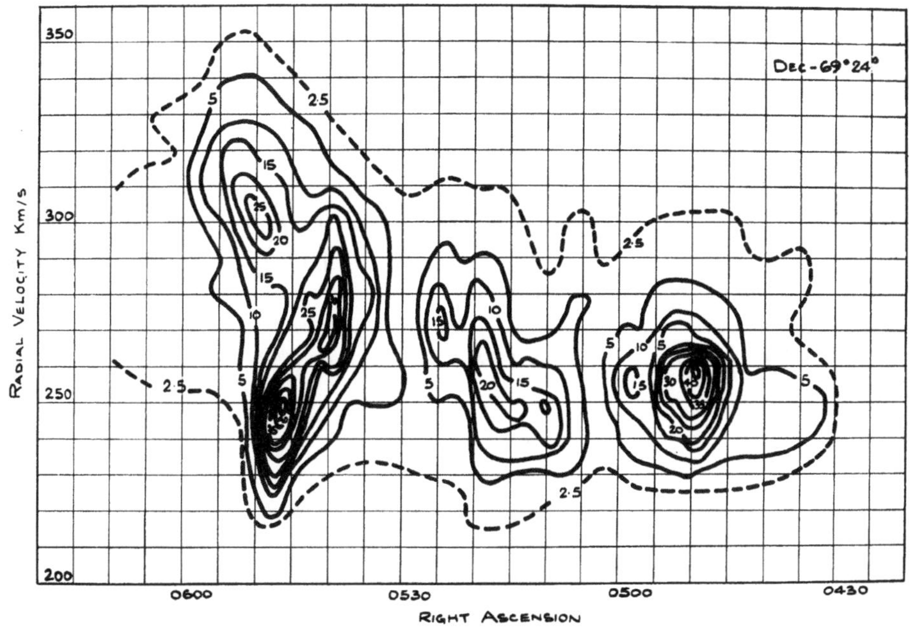

Fig. 4. Contours of relative intensity in the velocity-right ascension plane at declination −69°24′ (McGee and Milton, 1966).

Bok (1966) has brought together the mass estimates for Shapley Constellation I = NGC 1929-37 = Henize N44. The figures are $5.5 \times 10^6 M_{\odot}$ for the H I, $6 \times 10^4 M_{\odot}$ for the H II, with radio and optical measures agreeing well, and 3900 $M_{\odot}$ for supergiant stars with $M_v < -4.0$ .There is obviously a large amount of hydrogen available for star formation if the conditions are right. This type of comparison is obviously very difficult to make in the Galaxy, either for position, motion, or mass, as the galactic concentrations cannot be seen as separated entities in the same way. However, similar relationships probably exist in the Galaxy. At least we know that the major H I concentrations are about the same size in the two systems, the main difference being that the LMC concentrations are more self-contained because of the smaller external disrupting forces.

In the SMC, there is only a superficial correlation between the H I and H II (Hindman, 1967), in keeping with the presence of a fairly smooth H I distribution, but it should be noted that the H II regions are themselves generally weaker and smaller in the SMC.

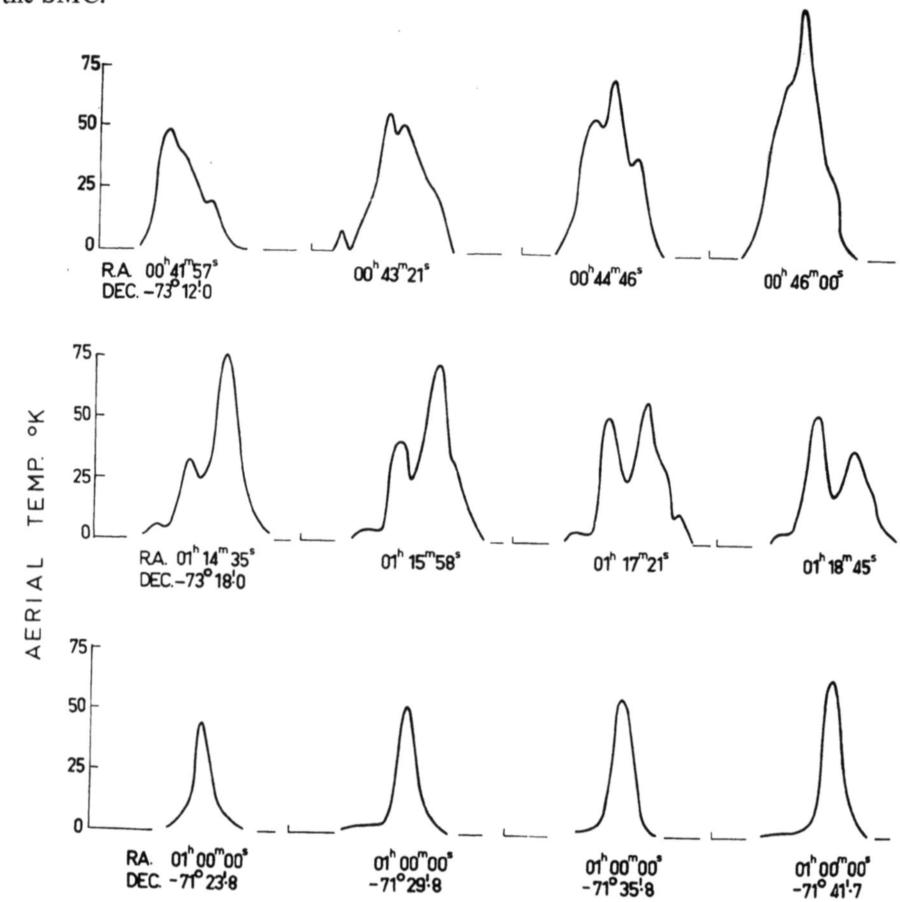

Fig. 5. An array of line profiles in the SMC (Hindman).

Relationships with clusters have been studied for both Clouds. In each case, no detailed correlation was found, but there is rough agreement in the boundaries of the region which contain the clusters and the main H I. A similar conclusion has been reported by Hindman for a comparison with cepheids in the SMC.

No correlation has been found with planetary nebulae in the LMC in either position or velocity, but studies by Feast and Webster indicate similar rotation parameters, although Webster suggests a different center of rotation. A striking result in any consideration of population II objects is that the optically-prominent bar in the LMC is not seen in the H I, except for some small H I concentrations in a few particular positions where H II regions are also found – these are probably in front of the bar, or behind it.

4. Details of Distribution and Possible Models

Both Clouds are clearly complex and irregular bodies, and we should not expect too definite an interpretation for the structure of either one. The situation is perhaps clearer for the SMC, and so we should consider it first.

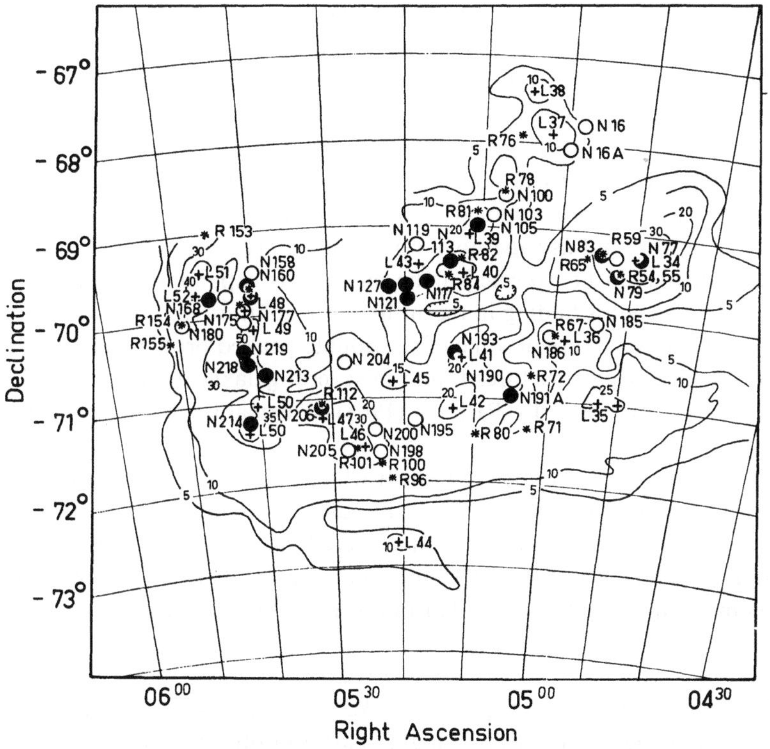

Fig. 6. Population I components of a section of the LMC. Some neutral hydrogen contours are included. + positions of H I complexes; ● positions of H II regions whose velocities are known; ○ positions of H II regions of size exceeding 2′ by 2′, but with no velocity information (labelled with Henize numbers); * supergiant stars whose radial velocities are known (labelled with Radcliffe numbers) (McGee and Milton, 1966).

A. SMC

We saw that the profiles are double-peaked over much of the area, suggesting two bodies of gas at different distances. There are three such regions. For the main one of these, Hindman (1964) made a comparison between the velocities of the two H I peaks, and those of several stars and SMC interstellar lines, as measured by Feast *et al.* (1960). In six cases where the stellar and calcium velocities differ, there is reasonable agreement between these differing velocities and the two H I velocities, while the absorption line velocity is lower than the stellar velocity, as in Figure 7. Hence, the gas with higher velocity must be the more distant, and the two bodies of gas must be moving away from each other in the line of sight at 40–50 km/sec.

To consider the possible geometry in more detail, we can examine a set of maps in the velocity-RA plane for various declinations. Examples are shown in Figures 8 and 9. As there appears to be at least a partial relationship between velocity and distance, these diagrams are roughly equivalent to sections through the SMC. Samples of these diagrams indicate a ring-like distribution, and Hindman's suggestion from the whole set of data is that we are seeing an expanding shell.

He finds evidence for three shells, in the positions shown in Figure 10. Each of the postulated shells has a diameter of $1\overset{\circ}{.}0$ to $1\overset{\circ}{.}8$, an expansion velocity of 20–25 km/sec, a hydrogen mass of about 10^7 $M_\odot$, and a maximum density of 1 atom/cm^3. Despite their prominence, the three together contain only 7% of the total H I mass, so the smooth distribution contains most of the hydrogen. There is no correlation observed between the shells and any other optical or radio constituent, but their age might be 10^7 yr, which would be very old as supernova remnants go. The geometry is not completely regular, but this could be expected if the material is expanding into a non-uniform medium. Also, Turner is finding in recent higher-resolution observations that there is some finer velocity structure in some of the double-peaked profiles. The shells may be related to the large ring structures found in several galaxies by Hayward (1964) and the big explosion proposed by Rickard (1968) to explain velocity anomalies in the Perseus arm.

The SMC shows a clear velocity gradient which suggests rotation (Figure 11). The maximum gradient for velocities corrected for galactic rotation is in position angle 55°. The velocity distribution for a 1° band centered on this major axis shows a clear turnover. The inclination of the SMC appears to be about 70°. The region of highest brightness might indicate an end-on bar or a very turbulent nucleus.

B. LMC

The LMC is more complex, as the gas is broken up into so many constituent parts. It is clear that the system is basically a highly-flattened object, inclined to the plane of the sky at about 27°. It is more flattened than the SMC.

Rotation has been clearly seen in several investigations, both in the H I and in the stars, with a P.A. $\approx$ 170° and a velocity range $\approx \pm 35$ km/sec. All the population I results give a rotation center about 1° away from the optically-derived centroid. This has always been a puzzle, but the effect is presumably just a consequence of an irregular mass distribution. The center of rotation is basically a center of symmetry for the velocities in the far northern and southern regions, and it is reasonable to suggest that the effective center of mass for these outer regions is different from the center of mass which is 'seen' by material further in. Alternatively, projection effects can cause

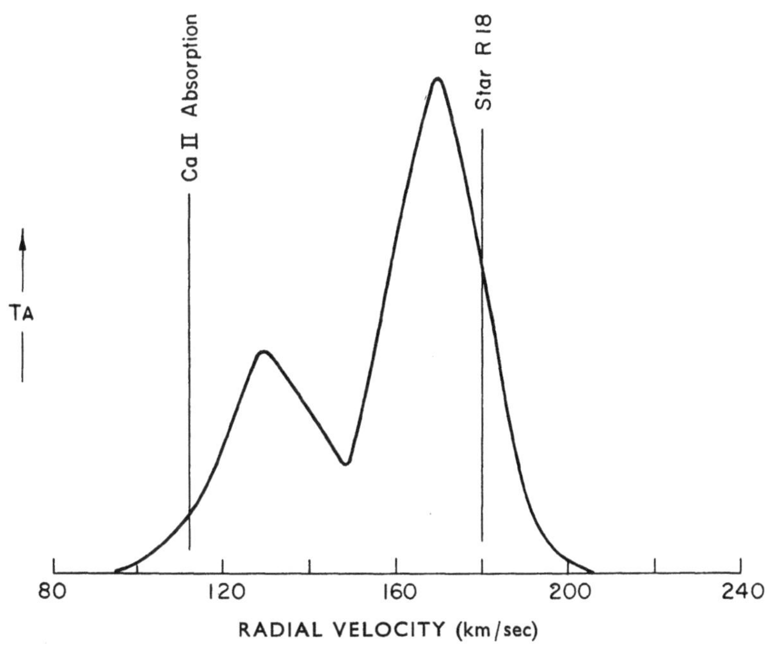

Fig. 7. Sample double-peaked profile at R.A. $00^h59.2^m$, Dec. $-72°22'$ (1975). The velocities of the star R 18 in the Radcliffe list and associated Ca II absorption are marked (Hindman, 1964).

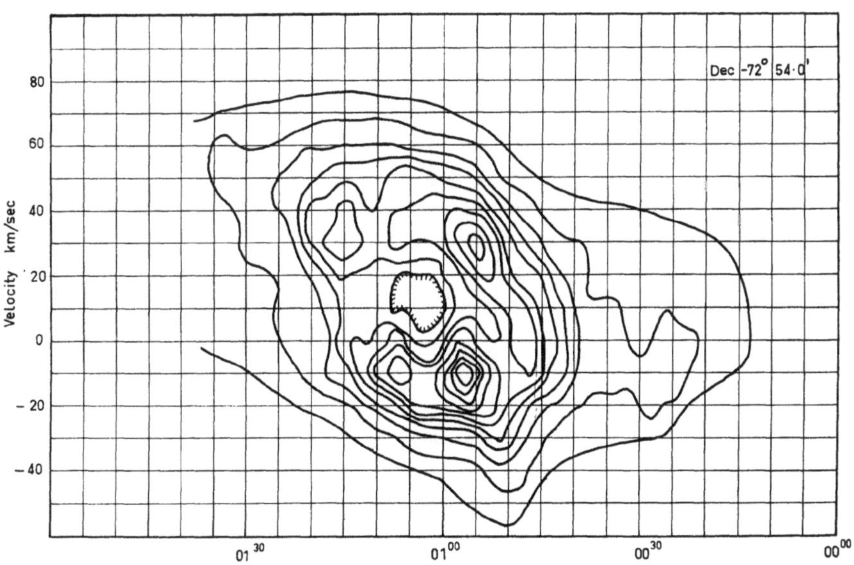

Fig. 8. R. A.-velocity diagram for a constant declination track in the SMC, showing the ring-like distribution of the central contours (Hindman, 1967).

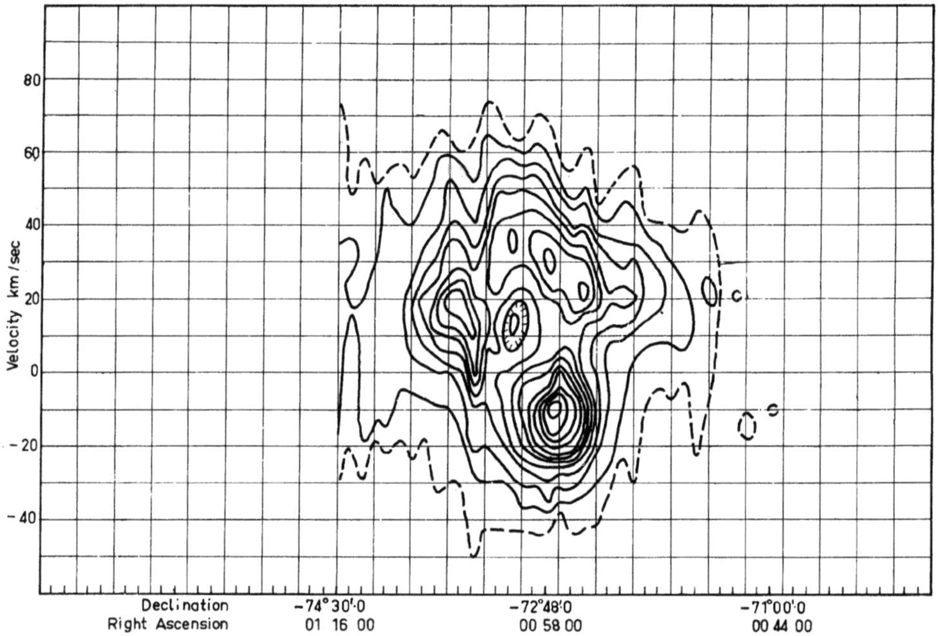

Fig. 9. Contours for a section through the SMC at right angles to the direction of maximum velocity gradient (Hindman, 1967).

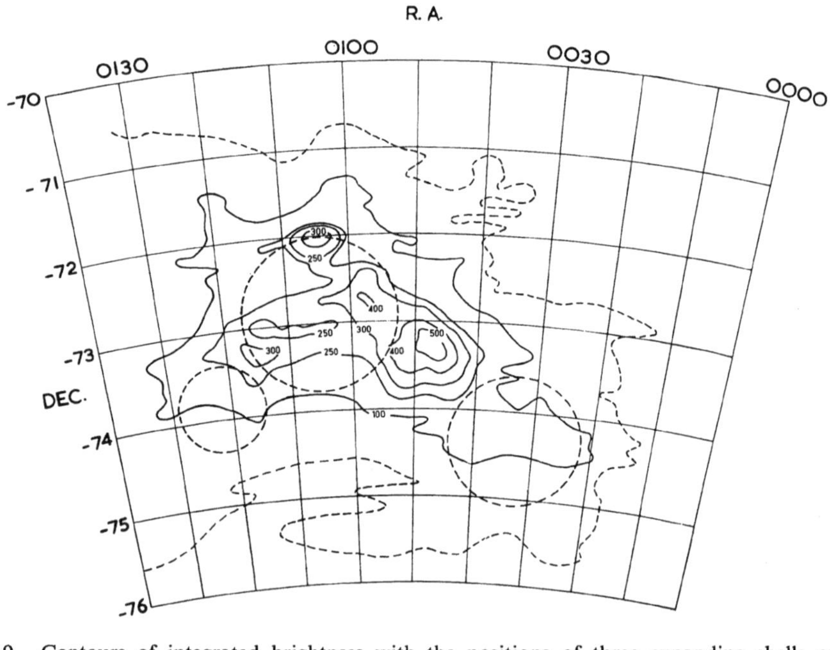

Fig. 10. Contours of integrated brightness with the positions of three expanding shells of gas indicated by dashed circles (Hindman, 1967).

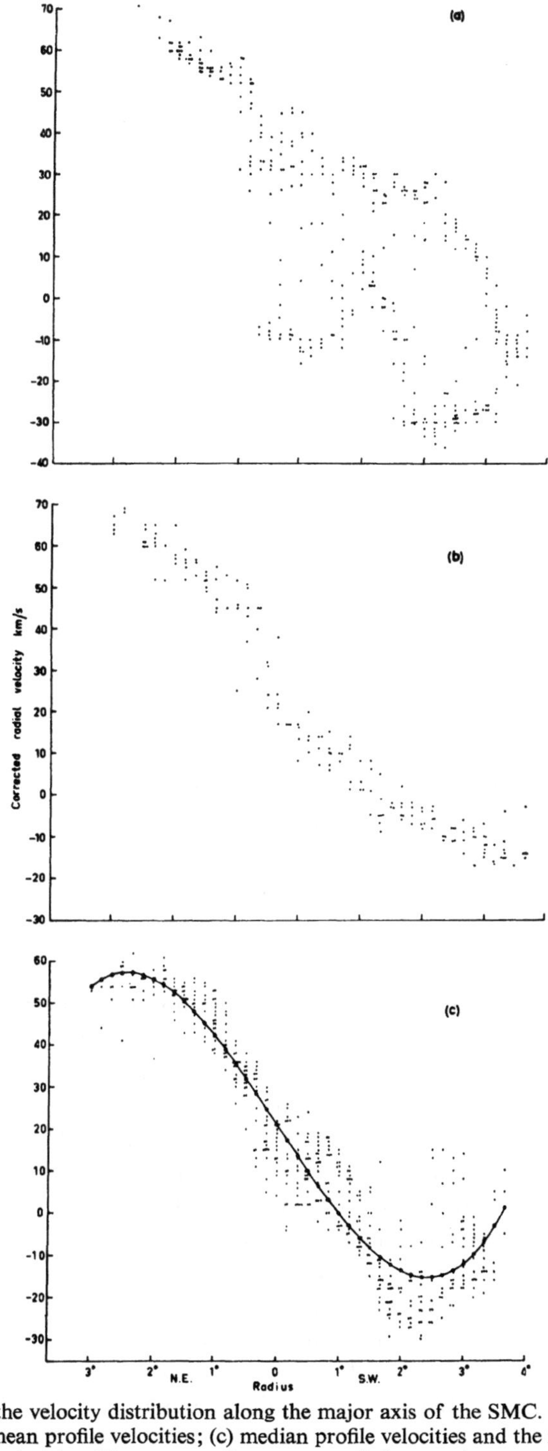

Fig. 11. Plots of the velocity distribution along the major axis of the SMC. (a) velocities from the profile peaks; (b) mean profile velocities; (c) median profile velocities and the rotation curve derived from the median velocities. The centre of symmetry of the rotation curve is at R. A. 01^h03^m, Dec. $-72°45'.0$. The points are from a band $\frac{1}{2}°$ on either side of the major axis, which lies in position angle 55° through the centre of the rotation curve (Hindman, 1967).

shifts in the apparent centers in a tilted system in which all constituents do not lie close to the same equatorial plane.

The presence of a rudimentary spiral pattern has been suggested by several people. McGee has gone furthest in this regard by proposing two sets of spiral arms, one of which lies in the main plane of the LMC, and the other is inclined to it an angle of about 20°. This model seems to me to attribute too much regularity to the system. It is based on a subdivision of the data into three velocity groups, which are considered

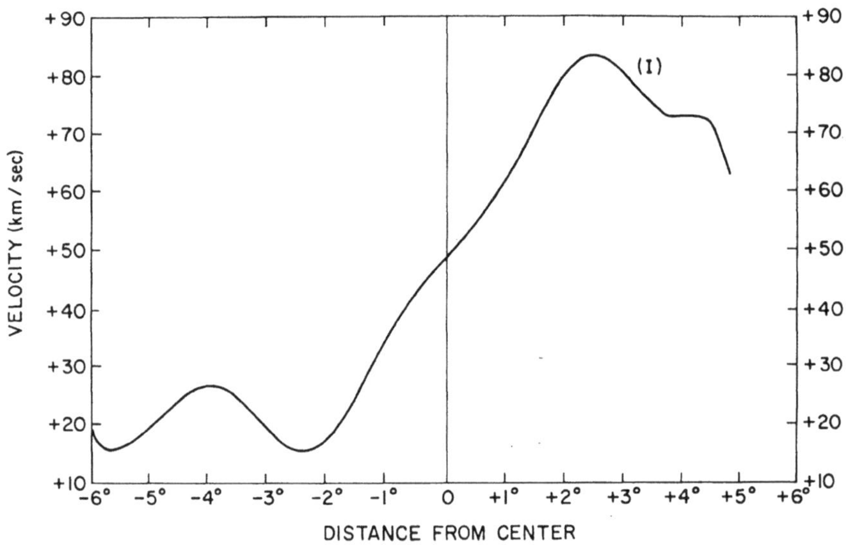

Fig. 12. Rotation curve for the LMC. The curve is a computer-fitted curve for all the points shown in Figure 13 (McGee and Milton, 1966).

to be largely independent of each other. These groups are centered on velocities of +300, +273 and +243. The first and third lead to the normal rotation pattern of material in the main disk, and the +273 group to a different rotation curve. Figure 12 shows the rotation curve for points near the major axis taking all three groups together, and Figure 13 shows separate rotation curves for the different groups. McGee suggests that the different slopes of the two curves imply two systems inclined to one another, but points out that this is not a unique solution.

The splitting of the hydrogen complexes into three groups of approximately equal H I mass is far from complete, however, as can be seen from the histogram in Figure 14 showing the velocity distribution of the H I in the listed complexes. Moreover, the total quantity of hydrogen in the complexes is only a third of that computed for the total hydrogen mass, and an integrated profile for all the H I in the LMC shows no sign of a splitting into velocity groupings. It therefore seems dangerous to base an LMC model on a clearcut separation of the gas into distinct groupings. Under these circumstances, the handling of the data becomes rather subjective, and different people might draw different conclusions. The two-plane model seems to me to require a more complete separation of the gas into distinct groupings than is justified by the data.

I would prefer to say merely that the gas in the LMC has a complex shape, with many localized concentrations, some at least of which are well away from the mean plane of the rather flattened system, and there may be a tendency in some regions

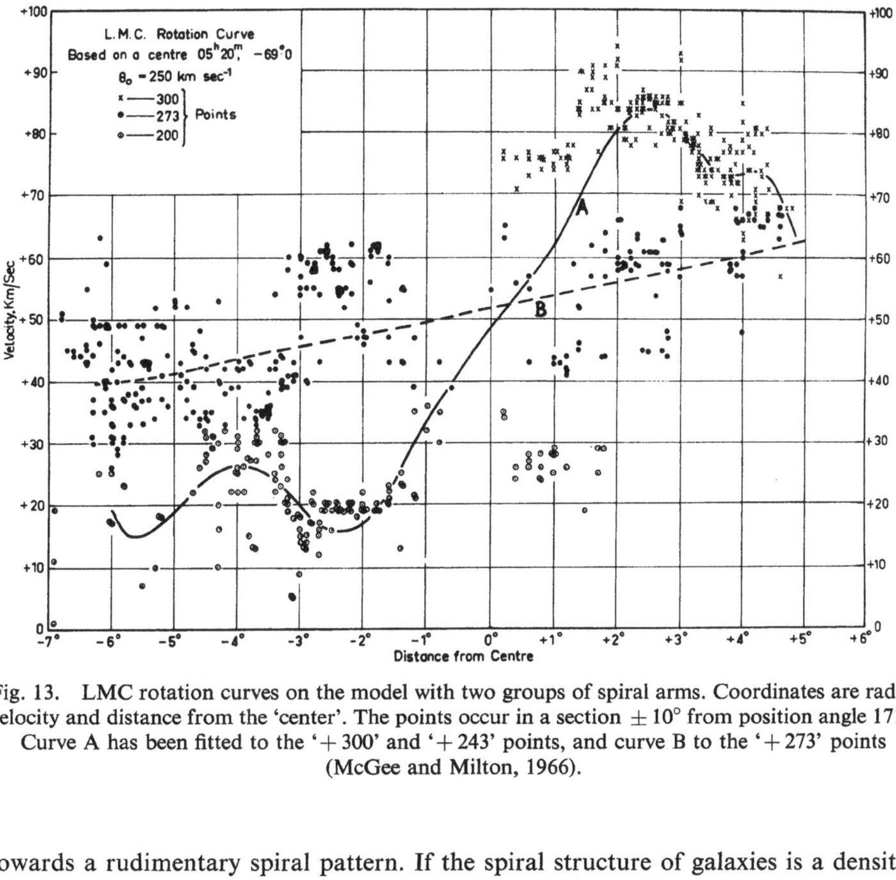

Fig. 13. LMC rotation curves on the model with two groups of spiral arms. Coordinates are radial velocity and distance from the 'center'. The points occur in a section $\pm 10°$ from position angle 171°. Curve A has been fitted to the '+300' and '+243' points, and curve B to the '+273' points (McGee and Milton, 1966).

towards a rudimentary spiral pattern. If the spiral structure of galaxies is a density-wave phenomenon, as discussed by Lin and others, we may speculate that the LMC is not sufficiently massive for a density wave to have become properly established. In the SMC, with still less mass, there is no sign of density wave fragments at all.

The values derived by McGee and Hindman for the total mass and hydrogen mass for the two Clouds are given in Table I, which clearly shows the great difference between the two Clouds in the relative proportions of gas.

The estimates of total mass are very uncertain, as is well known. It should also be stressed that the hydrogen masses should be regarded as lower limits. Recent evidence,

TABLE I

	$M_{TOT}(M_\odot)$	$M_H(M_\odot)$	M_H/M_T
LMC	6.1×10^9	5.4×10^8	0.09
SMC	1.5×10^9	4.8×10^8	0.32

particularly from observations in the Galaxy, suggest that hydrogen self-absorption is probably more important than previously thought, due to complexities of the small-scale cloud structure. The gas masses could easily be double the earlier estimates.

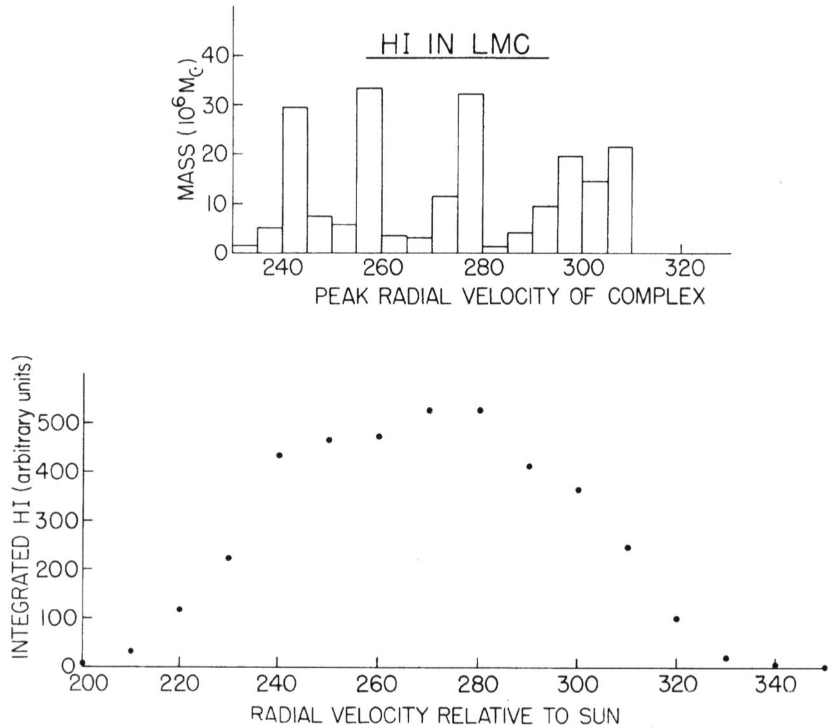

Fig. 14. (a) Histogram of HI mass in LMC complexes listed by McGee and Milton (1966). (b) Integrated hydrogen for whole LMC, as a function of velocity (from constant-declination contour diagrams).

5. The Bridge

Turner has a paper almost ready for publication, which is based on observations of the bridge region with the Argentine 30-meter telescope, operated jointly by the Carnegie Institution of Washington and the Argentine Institute of Radioastronomy. His map for the integrated hydrogen (Figure 15) shows that the bridge is continuous

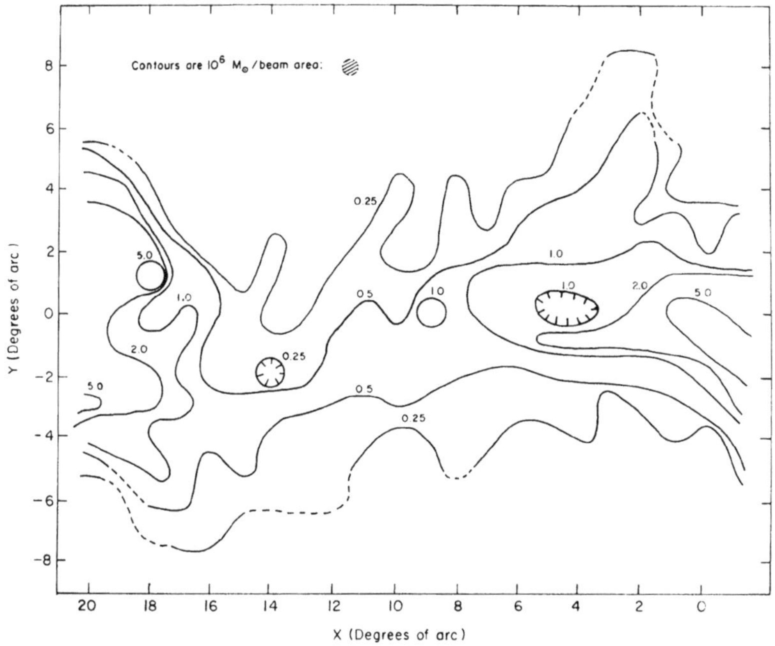

Fig. 15. Integrated brightness in the 'bridge' between the Clouds (Turner, to be published).

on the sky and also quite narrow. However it seems to be composed of about 20 fairly separate units, as can be partially seen from a diagram for the hydrogen in one specific velocity range (Figure 16). These separate units have velocity halfwidths of 40–80 km/sec, and the total mass contained in all of them is about 10% of the hydrogen mass of the whole Magellanic System. Some of the subunits resemble small scale concentrations of the LMC type, others might be similar to the objects we know as high velocity clouds in galactic observations. Some in fact show velocities up to 60 km/sec different from the main motion and might be considered as high velocity clouds in the Magellanic frame of reference.

Turner's work indicates particular regions of high hydrogen density where it would be most appropriate to look for stars in the bridge. One of these is an unresolved feature situated 9° from the SMC and containing 10^6 $M_\odot$ of hydrogen. Two 'fingers' can be seen emerging at apparently different levels from the SMC.

Hindman has also found some gas concentrations and a possible ring in the bridge region, but he has only covered limited sections so far.

6. The System

The LMC, SMC and the bridge should clearly be regarded as forming one system. The fact that the outer edges of the two Clouds are fairly sharp, while there is a stream of material between the Clouds, suggests that the LMC and SMC may be

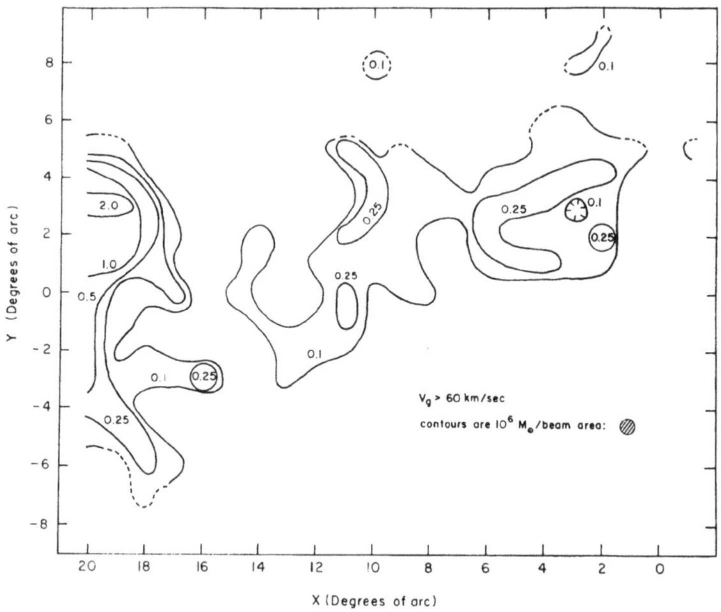

Fig. 16. Integrated brightness in the bridge, for V_g (velocity corrected to a stationary Galaxy) > 60 km/sec (Turner, to be published).

separating. We cannot of course measure the transverse velocity, but the time scale of the separation might be of the order of 5×10^8 yr, if we make an estimate from the observed radial velocities. It is interesting for comparison that Hunter and Toomre (1969a) propose a similar time period as a possible time that has elapsed since the Clouds last passed through the outer galactic disk if they are in fact satellites of the Galaxy.

If the two Clouds are coupled together through the bridge there are interesting dynamical problems for the theorists, when we consider that each Cloud appears to be rotating separately while at the same time they have a loose junction between them. Similar problems are of course presented by other pairs or groups of galaxies with bridges. One often wonders how such systems can be stable.

A possible link between the Clouds and the Galaxy has been looked for in Australia and more systematically in Argentina. Varsavsky has found material extending from the Galaxy part of the way towards the Small Cloud, but a complete bridge has not yet been seen. Effects produced by the Magellanic Clouds on the gas in the Galaxy are of course known, but this subject is too extensive to be discussed here.

7. Other Radio Lines

Radhakrishnan (1967) has looked for OH emission in several likely places in the LMC, but without success. McGee, Batchelor and Brooks are at present surveying H II regions in the LMC for the H 109 α recombination line near 6 cm wavelength.

8. Conclusion

The most interesting results of studies of the neutral hydrogen in the Clouds concern the interrelationships between the various population I constituents and also the internal structure and motions of the system. More detailed studies could give more insight into the structure of each Cloud and the bridge, and also there are problems connected with the dynamics of the whole system.

References

Bok, B. J.: 1966, *Ann. Rev. Astron. Astrophys.* **4**, 95–144.
Feast, M. W., Thackeray, A. D., and Wesselink, A. J.: 1960, *Monthly Notices Roy. Astron. Soc.* **121**, 337–85.
Hayward, R.: 1964, *Publ. Astron. Soc. Pacific* **76**, 35.
Hindman, J. V., Kerr, F. J., and McGee, R. X.: 1963, *Australian J. Phys.* **16**, 570–583.
Hindman, J. V.: 1964, *Nature* **202**, 377–378.
Hindman, J. V.: 1967, *Australian J. Phys.* **20**, 147–171.
Kerr, F. J., Hindman, J. V., and Robinson, B. J.: 1954, *Australian J. Phys.* **7**, 297–314.
McGee, R. X.: 1964, *Australian J. Phys.* **17**, 515–523.
McGee, R. X. and Milton, J. A.: 1966, *Australian J. Phys.* **19**, 343–374.
Radhakrishnan, T.: 1967, *Australian J. Phys.* **20**, 203–204.
Rickard, J. J.: 1968, *Astrophys. J.* **152**, 1019–1042.

PART II

CURRENT OBSERVATIONAL APPROACHES

INTERMEDIATE BAND PHOTOMETRY OF THE BRIGHTEST STARS IN THE MAGELLANIC CLOUDS

EUGENIO E. MENDOZA V.

Instituto de Astronomía, Universidad de México
and Departamento de Astronomía, Universidad de Chile

During the past three years we have occasionally observed stars located all over the sky in the (33, 35, 37, 40, 45, 52, 58, 63)-photometric system which has been defined by Johnson *et al.*, (1967), at the Lunar and Planetary Laboratory, La Silla, European Southern Observatory, and Cerro Tololo Inter-American Observatory.

A comparison of the observed atmospheric extinction is given in Table I. The columns of this table contain, first, the filter band; second, the effective wavelength (see Johnson *et al.*, 1967); third through last, the observed atmospheric extinction at Catalina, Tololo, and La Silla observatories, respectively. Although Tololo seems to have the smallest extinction coefficients these three observatories give the same results, for practical purposes.

This paper shows the intermediate band photometry of 36 objects that belong either to one or another of the two Magellanic Clouds, obtained at Tololo in November, 1968, in the (33, 35, 37, 40, 45, 52, 58)-system. The results of these observations are given in Tables IIa and IIb. The columns of these tables contain, first, the Radcliffe serial number, *R* (Feast *et al.*, 1960); second, the spectral type given by Feast *et al.* (1960); third, the 52 magnitude; fourth through last, the color indices (33–35), (35–37), (37–40), (40–45), (45–52), and (52–58), respectively. The stars were observed on one or two different nights. The probable error of a single observation (in our photometry) is approximately 0.015 mag.

TABLE I

Observed atmospheric extinction

Filter band	$\lambda_0(\mu)$	*k*				
		Catalina 1965–66	Tololo 1966 Feb.	Tololo 1966 May	Tololo 1968 Nov.	La Silla 1968 Jan.
33	0.337	0.692	0.643	0.671	0.635	0.686
35	0.353	0.570	0.544	0.562	0.529	0.587
37	0.375	0.459	0.409	0.464	0.423	0.463
40	0.402	0.352	0.303	0.341	0.316	0.364
45	0.459	0.237	0.192	0.220	0.199	0.241
52	0.518	0.180	0.162	0.149	0.139	0.173
58	0.583	0.164	0.141	0.108	0.117	0.163
63	0.635	0.122	0.097	0.064	–	0.150

Muller (ed.), The Magellanic Clouds, 69–73.

TABLE IIa

Observational data Small Magellanic Cloud stars

R	Sp.	52	33–35	35–37	37–40	40–45	45–52	52–58
5	B3 I	11.01	−0.07	−0.36	−0.61	−0.12	0.04	0.03
6	B6 I	11.06	−0.01	−0.33	−0.56	−0.01	0.09	0.07
8	A0 Ia:	11.24	0.01	−0.48	−0.77	−0.05	0.09	0.04
9	B3 Ia:	11.12	−0.12	−0.55	−0.66	−0.13	0.02	0.01
10	A0 Ia:	10.97	0.10	0.25	−0.39	0.04	0.14	0.09
11	B6 Ia	10.81	−0.02	−0.37	−0.58	−0.03	0.28	0.05
16	B	12.54	−0.03	−0.05	−0.45	−0.03	0.06	0.10
18	B0.5 I	12.07	−0.12	−0.58	−0.60	−0.17	−0.01	0.04
19	A3 Ia	11.62	0.15	0.64	−0.18	0.03	0.11	0.08
21	A0 Ia:	11.42	−0.01	−0.02	−0.48	−0.05	0.05	0.04
22	A3 I	12.33	0.14	0.59	0.02	0.03	0.10	0.03
23	A1 I	12.11	−0.05	−0.24	−0.39	−0.09	0.07	−0.09
24	A0:	12.55	0.32	0.43	−0.26	−0.19	0.14	−0.03
25	B1:	12.81	−0.18	−0.46	−0.61	−0.22	0.02	−0.19
27	B9 Ia	10.90	0.00	0.05	−0.50	−0.05	0.08	0.06
28	B0 I	13.94	−0.17	−0.61	−0.67	−0.20	−0.27	0.20
31	O: f:	12.28	−0.28	−0.14	−0.78	−0.10	−0.10	−0.07
32	B	12.72	−0.18	−0.34	−0.62	−0.09	−0.01	0.01
33	B	12.67	−0.11	−0.56	−0.62	−0.15	0.02	−0.03
35[a]	B1 I:	12.59	−0.09	−0.27	−0.51	−0.12	−0.03	−0.03
36	B3 I	11.34	−0.05	−0.41	−0.62	−0.09	0.02	−0.01
37	B6 I	11.28	−0.05	−0.25	−0.58	−0.04	0.06	0.04
42	B2.5 I	11.04	−0.19	−0.40	−0.66	−0.13	0.00	−0.01

[a] A faint companion included in the diaphragm.

TABLE IIb

Observational data Large Magellanic Cloud stars

R	Sp	52	33–35	35–37	37–40	40–45	45–52	52–58
51	B1.5 Ia:	11.37	−0.05	−0.51	−0.64	−0.06	0.09	0.06
52	B8 I	10.56	−0.01	−0.27	−0.54	0.00	0.11	0.10
53	B0 Ia	11.36	−0.14	−0.58	−0.67	−0.17	−0.03	−0.04
56	B2 Ia	11.67	−0.09	−0.50	−0.62	−0.19	0.01	−0.04
57	B3 I	10.83	−0.07	−0.45	−0.64	−0.11	−0.01	0.00
58	B5 I	11.08	−0.06	−0.08	−0.56	−0.08	0.11	0.07
61	A0 Ia	10.85	−0.03	−0.21	−0.54	−0.06	0.04	0.01
68	B8 I	11.02	−0.02	−0.23	−0.54	−0.07	0.03	0.00
70	B3 I	11.21	−0.08	−0.39	−0.58	−0.14	0.04	−0.01
72	A0 Ia	11.72	−0.06	−0.42	−0.60	−0.12	0.03	0.01
73	B8 Ia	10.62	−0.07	−0.33	−0.60	−0.09	0.05	0.03
75	A0 Ia	10.40	0.03	−0.16	−0.54	−0.03	0.10	0.07
79	B5 I	12.08	−0.10	−0.40	−0.57	−0.09	0.01	0.06

The seven-color photometry analysis presented here could also include a larger number of comparisons between the several color indices. Higher members of the Balmer series are crowded in Filter 37's transmission region making the color index (37–40) very sensitive to luminosity variations. Miss Divan (1954) has shown that the difference between the intrinsic gradients in ultraviolet and in blue, is sensitive to the spectral type in the region of O and B stars.

Figure 1 shows the relation between (40–45) and (33–35), left side; and that of (40–45) and (37–40), right side. The scatter shown in this last relationship is probably because of differences of reddening and luminosity among the plotted stars. On the left side we note that O and B galactic supergiant stars are separated from the Magellanic Cloud stars, due perhaps to the latter's ultra-violet deficiency when compared with the former, because of evolution or chemical composition possible differences between the O and B galactic supergiants and the stars listed in Table II.

The photometric system used in this investigation is similar to the seven-color system that Borgman (1960, 1961) defines. In order to use Borgman and Blaauw's (1964) β and δ parameters, our data must be changed into Borgman's system. Eighteen galactic supergiants, between 09.5 and B 9, observed on both Johnson *et al.*, and Borgman and Blaauw's photometric systems were used when transforming these equations into Borgman and Blaauw's system (1964). The resulting equations are given in Table III. By the way, the galactic supergiant stars plotted in Figure 1 were taken from the same stars used to find the data listed in Table III.

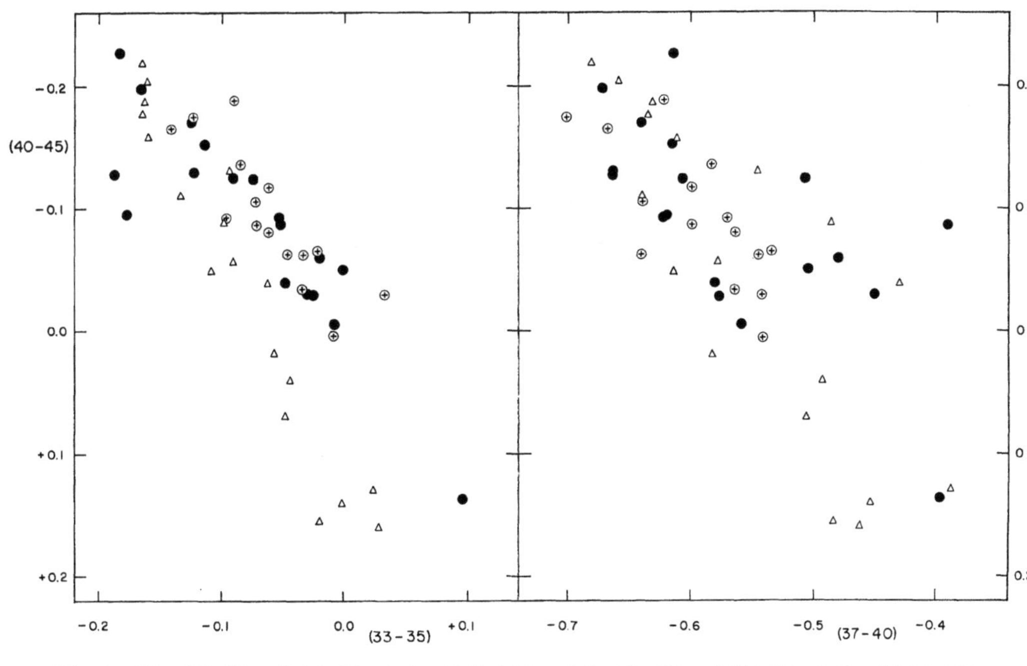

Fig. 1. The (40–45) and (33–35) relation (left side) and the (40–45) and (37–40) relation (right side). See text for an explanation of the symbols.

TABLE III

Transformation equations to Borgman and Blaauw's system (from early type supergiant stars only)

$N-M=0.72(40\text{–}45)+0.172$	(± 0.002 mag)
$\beta=1.20(37\text{–}40)-0.50(40\text{–}45)+0.695$	(± 0.005 mag)
$\delta=2.15(33\text{–}35)-0.73(40\text{–}45)+0.205$	(± 0.006 mag)

Figure 2 illustrates graphically the relation between β and δ for stars in the Magellanic Clouds. In this figure solid and broken lines have been taken from Borgman and Blaauw's calibrated β, δ-diagram (1964, Figure 8). These are constant absolute magnitude and intrinsic color lines, respectively; filled and crossed circles represent Small and Large Magellanic Cloud stars, respectively. The two shadowed lines on top could be constant absolute magnitude lines, perhaps brighter than -8.0 mag.

Supposing that the position of the Magellanic Cloud stars in this β, δ-diagram is only due to luminosity effects, we notice that almost all the stars listed in Table II are more luminous than the most luminous galactic O and B supergiants. In Figure 2 all Large Magellanic Cloud stars observed, are above the line $M_v=-7$. Their range in the V magnitude from 10.5 to 12.0 mag. Since the scatter of these stars in the β, δ-plot is not too large (see also Figure 1) we assume that all of them have a constant luminosity, for instance, $M_v=-8.0$ to obtain a distance modulus, corrected for interstellar extinction, equal to 18.9 ± 0.2 which is in perfect agreement with other

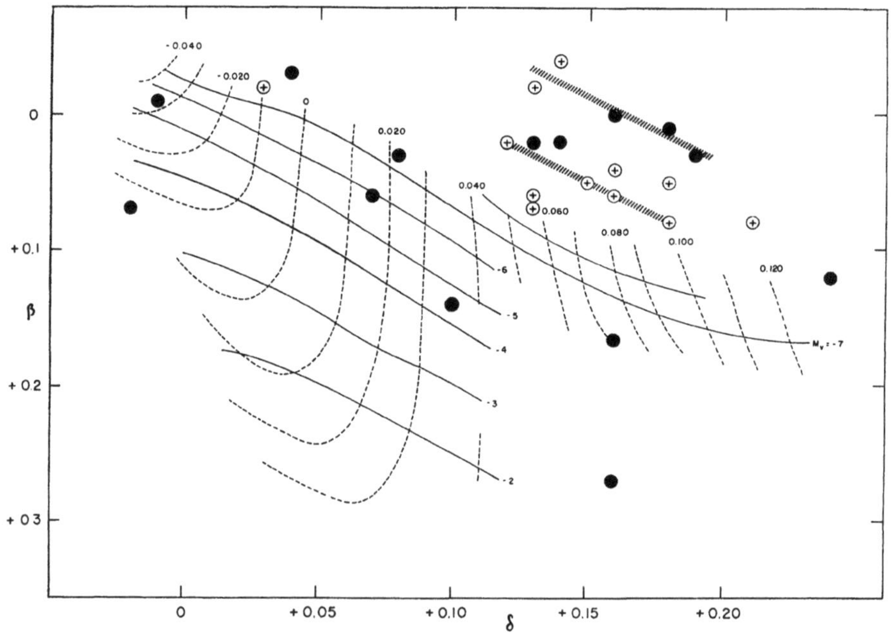

Fig. 2. The calibrated β, δ-diagram (see text).

determinations (Bok 1966). For this computation, we have assumed a ratio between total absorption and color excess $A_v/E_{N-M}=8.4$ valid for h and χ Persei, following Borgman and Blaauw's (1964). The V magnitude was obtained from the equation $V=(52)-0.0059-0.5362(52-58)$ (see Johnson *et al.*, 1967). The intrinsic $(N-M)_0$ color, from Figure 2. Because the stars listed in Table II have a small reddening, the precise knowledge of the ratio $A_v/(N-M)$, is not very important.

It should be pointed out, that the distance modulus of the Small Magellanic Cloud derived from the calibrated β, δ-diagram turns out to be smaller than that of the Large Magellanic Cloud. The difference could be as high as 0.5 mag.

We also have made *BVRI* photometric observations of 100 objects that belong to the Magellanic Clouds (Mendoza, 1970). They indicate that again some of the Magellanic Clouds stars are more luminous than the most luminous galactic supergiants and they also have a color excess in $V-R$ unexplained by interstellar extinction alone. There are many possible explanations for the additional flux in the red wavelength such as: (a) emission line radiation in the object's neighbourhood; (b) free-free radiation in the stellar vicinity; (c) a peculiar extinction law; (d) incorrect intrinsic colors; (e) anomalous stellar emission; (f) late type companions; (g) circumstellar dust clouds; and (h) different atmospheric chemical composition between stars in the Clouds and in the Galaxy. Either (g) or (h) or both are the most likely (cf. Mendoza, 1970) causes for the $V-R$ color excess.

It is difficult to apply the photometric method (Mendoza, 1967) to derive the distance of the Clouds. A comparison of the Hertzprung-Russell diagram for the Large Magellanic Cloud and the Double cluster, h and χ Persei indicates that the distance modulus of LMC should be larger than 18.0 mag. (cf. Mendoza, 1970). The derived difference in distance moduli of the clouds is approximately 0.25 mag. LMC being closer than SMC by this amount.

Acknowledgement

Telescope time was granted according to an agreement between AURA, Inc. and the University of Chile. We are grateful to Dr. Blanco for the many facilities given to us during the time we stayed in Tololo.

References

Bok, B. J.: 1966, *Ann. Rev. Astron. Astrophys.* (ed. by L. Goldberg), Annual Reviews, Inc. Palo Alto, Calif., U.S.A. **4**, 95.
Borgman, J.: 1960, *Bull. Astron. Inst. Neth.* **15**, 255.
Borgman, J.: 1961, *Bull. Astron. Inst. Neth.* **16**, 99.
Borgman, J. and Blaauw, A.: 1964, *Bull. Astron. Inst. Neth.* **17**, 358.
Divan, L.: 1954, *Ann. Astrophys.* **17**, 456.
Feast, M. W., Thackeray, A. D., and Wesselink, A. J.: 1960, *Monthly Notices Roy. Astron. Soc.* **121**, 337.
Johnson, H. L., Mitchell, R. I., and Latham, A. S.: 1967, *Comm. LPL* **6**, 85.
Mendoza, E. E.: 1967, *Bol. Obs. Tonantzintla and Tacubaya* **4**, 149.
Mendoza, E. E.: 1970, *Bol. Obs. Tonantzintla and Tacubaya* **5**, 269.

A SEARCH FOR RED VARIABLE STARS IN THE MAGELLANIC CLOUDS

T. LLOYD EVANS
Radcliffe Observatory, Pretoria, South Africa

1. Introduction

A systematic study of red variable stars in the Magellanic Clouds is of interest in connection with the older stellar populations. Two fields in each Magellanic Cloud, as well as two comparison fields, are being photographed with the 74-inch (1.88 m) reflector.

TABLE I

Field	α	δ
	(1900)	
NGC 371 (SMC)	$01^h00^m.2$	$-72°49'$
Radcliffe Variable Star Field (SMC)	00 48.3	−73 38
HD 33617 (LMC)	05 06.2	−68 33
S Doradus (LMC)	05 18.8	−69 21
SMC Comparison Field (Galaxy)	22 00.5	−67 00
LMC Dessy Variable Star Field 'A'	05 39.0	−70 18

In each Cloud the first field is relatively open and the second very dense. Photoelectric *B*, *V* sequences are available for the first 3 fields. Each plate has a field of $40' \times 30'$.

2. Observations

A V plate ($103aD+GG\ 11, 30^m$) is taken at every possible dark of moon. Infrared plates ($I-N+RG8, 60^m$) and B plates ($103aO+GG13,\ 30^m$) have been taken occasionally. Red stars are selected primarily by their large $V-I$ index, while the more extreme C stars may be distinguished from M stars by their faintness in B. Highly reddened stars of earlier type may also be included in the sample thus isolated.

The red stars are subdivided by $V-I$ colour class as follows:

(1) Image strength on I equal or slightly less than on V.

(2) Image strength on I slightly greater than on V.

(3) Image strength on I considerably greater than on V.

(4) Image strength on I very much greater than on V.

Observations of the NGC 6522 field near the galactic centre show a close corre-

Muller (ed.), The Magellanic Clouds, 74–78.

lation between colour class and M subtype, as expected from the work of Blanco (1964). Photo-electric sequences in some of the fields will be used to check on possible variation in the colour class boundaries with magnitude.

3. Blink Searches for Variable Stars

Two independent blink searches are being made:

(i) A general blink of all stars on V plates, the infrared plates being used subsequently to weed out 'blue' variables.

(ii) A more careful blink of stars already known to be red, in which smaller amplitude variables may be detected.

The partial results now available suggest a trend in the order Galactic Centre, LMC, SMC in decreasing proportion of late M stars and increasing frequency of variables of comparatively large amplitude among stars of smaller V – I. This may be caused by the increasingly severe selection by absolute magnitude, coupled with a higher proportion of carbon stars (which have smaller V – I than M stars) in the Magellanic Clouds.

4. Comparison with Other Surveys

The recovery rate of Harvard variables has been poor, especially in the S Doradus field where only 5 (none red) were detected in blinking 10 plate pairs. Many are too bright for effective blinking on our V plates.

A V – I pair of the Dessy Variable Star Field 'A' in the LMC (Dessy and Laborde, 1966) showed that only 9 out of 148 Cordoba variables are red by our definition. Most red stars would be very faint or invisible on the blue plates used by Dessy; on the other hand we have found many red variables in the NGC 6522 field which were not detected by Gaposchkin (1955) on blue plates. The rapid increase in $V-I$ and levelling off of $B-V$ as TiO bands strengthen means that there is no conflict with Dessy's description of many Cordoba variables as red.

There are many red stars in Dessy Field 'A' which in some cases may result from the heavy obscuration apparent in this field.

5. Large Amplitude Variables

We regard as a Mira any red variable with an amplitude of several magnitudes or which is reasonably regular and of long period with minimum below the plate limit. To date the S Doradus field has yielded 12 such stars with $126^d < P < 420^d$; those with $P \sim 200^d$ are the brightest, $V_{max} \sim 16$. A less complete survey in the Radcliffe Variable Star Field in the SMC has detected 4 stars with $240^d : < P < 360 :^d$, all of them rather faint. The latter field contains in addition 8 much brighter stars with $300^d < P < 700^d$, several of which were discovered earlier by Harvard or Cordoba.

The stars of the first group are considered to be true Mira variables and follow a

$P-L$ relation like those in the Galaxy, with the longer period stars being faintest at maximum. They are evidently quite common in the LMC, where the strong representation of those with P near 200 days suggests that the LMC has a population similar to that typified by the metal rich globular clusters in the Galaxy.

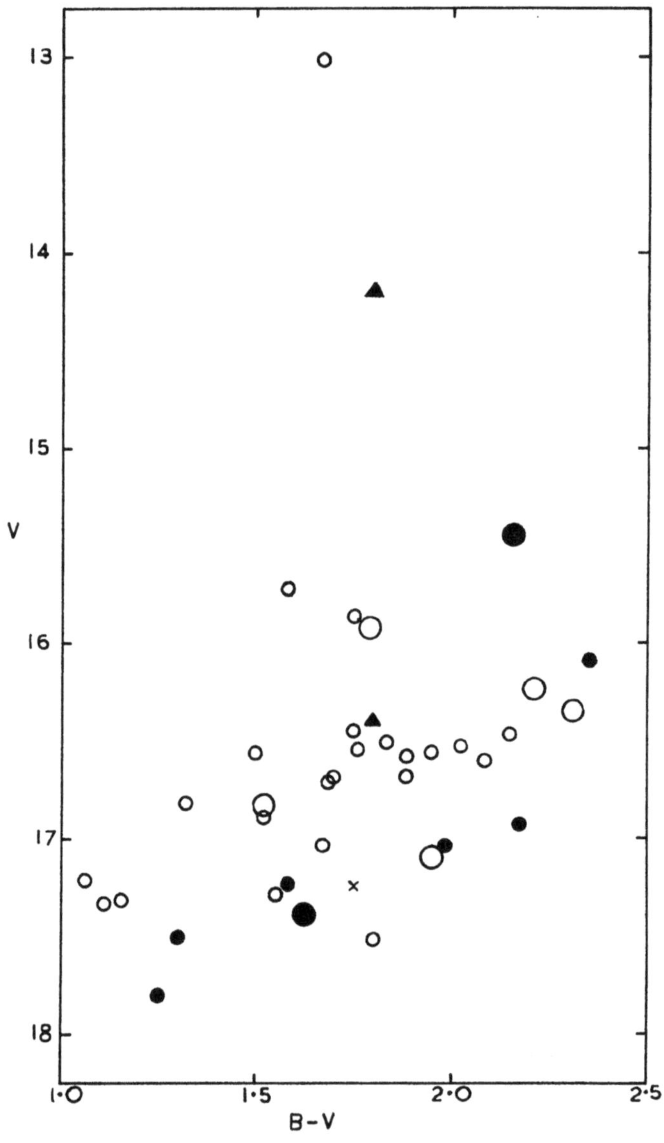

Fig. 1. The V, $B-V$ diagram for red stars in the NGC 371 field. Open circles, filled circles and crosses correspond to $V-I$ colour classes 1, 2 and 3 respectively; large symbols denote the instantaneous positions of variable stars. The triangles represent extreme positions of HV 12179. Six stars with $B-V< +1.00$, perhaps because they have blue companions, are omitted. The reddening $E_{B-V}=0.10$ estimated by Andrews has not been allowed for.

The second group comprises the bright long period variables found in the SMC by Harvard workers. These are clearly separated in magnitude from the true Miras for $P \gtrsim 300^d$. Those in our field have magnitudes at maximum in the range $V = 12 - 15$ with amplitude $\Delta V > 5^m$ in some cases. Those of longer period are generally brighter in accordance with the results of Payne-Gaposchkin and Gaposchkin (1966). They are presumably supergiants. Spectroscopic observations are being made at maximum.

6. NGC 371 Field

55 red stars were found, of which 39 belong to $V-I$ colour class 1. The dividing line between 1 and 2 corresponds to $V-I \sim 2.1$, or spectral type M 2–3 (Blanco, 1964).

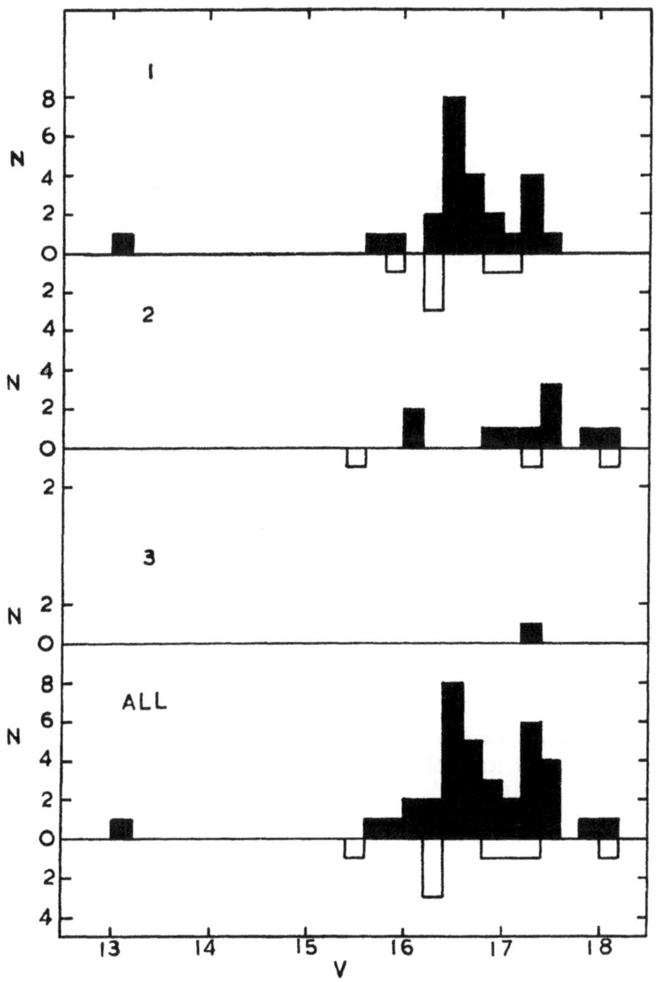

Fig. 2. The distribution with V magnitude of variable (below the line) and non-variable red stars in the NGC 371 field.

Only 1 star belongs to colour class 4. The photo electric sequence of P. J. Andrews has been used to calibrate photographic photometry of 44 stars in a field of 24′ diameter. The colour magnitude diagram is shown in Figure 1 and histograms of apparent magnitude in Figure 2. The SMC Comparison Field contains 13 red stars, all of $V-I$ colour class 1; these do not show the concentration to $V\sim 16.5$ of the SMC stars. It seems probable that the sample we have isolated, rising to $M_v=-2.9$, comprises stars from the red giant tip of an old stellar population (Arp, 1958). The stars with $B-V\sim 2.0$ are often regarded as being possible carbon stars; it is interesting to note that the cutoff near $B-V=2.2$ is similar to that found for galactic S stars which are not also large amplitude variables.

Seven plate pairs were blinked, revealing 6 certain variables. These include HV 12179 which has a range of 14.2 to 16.4 in V and a star which was seen for at least 6 months in 1968 but not in earlier or later seasons, with a flat maximum at $V\sim 18$ in August, 1968.

7. Summary

A progress report on a search for red variables in the Magellanic Clouds includes the discovery of several Mira variables, with magnitudes at maximum of $V\sim 16$ and fainter. Several new examples of the much brighter Shapley long-period variables have been found in the SMC. Many variables with smaller amplitude have been found; these are being studied in order to determine the incidence of variability with colour and magnitude. A colour magnitude diagram of red stars found near NGC 371 in the SMC shows the well-known red giant region from $V=16$ to 17.5, extending to $B-V\simeq 2.2$. Several of these stars are small amplitude variables.

Acknowledgements

I am indebted to Dr. A. D. Thackeray for initiating this project, to Dr. M. W. Feast for discussions and to Dr. P. J. Andrews and to Dr. A. J. Wesselink for the photoelectric sequences near NGC 371 and the Radcliffe Variable Star Field, respectively.

References

Arp, H.: 1958, *Astron. J.* **63**, 273.
Blanco, V. M.: 1964, *Astron. J.* **69**, 730.
Dessy, J. Landi and Laborde, J. R.: 1966, *Bol. Inst. Mat. Astron. Fis.* **2**, 3.
Gaposchkin, S.: 1956, *Per. Zv.* **10**, 337.
Payne-Gaposchkin, C. and Gaposchkin, S.: 1966, *Smithsonian Contr. Astrophys.* **9**.

PRELIMINARY RESULTS OF A PHOTOMETRIC STUDY OF THE NGC 371 REGION IN THE SMC

P. J. ANDREWS
Radcliffe Observatory, Pretoria, South Africa

The region of the SMC about NGC 371 was that chosen by Arp (1960) to investigate the Period-luminosity relation for cepheids in the SMC. Arp's work was based on photographic photometry of the cepheids calibrated with a photoelectric sequence measured on the Radcliffe 74-in. (1.88 m) reflector. The period-luminosity relation found by Arp is in poor agreement with those found by others for the SMC (Gascoigne and Kron, 1965) and the LMC (Woolley *et al.*, 1962). One possible reason for this disagreement is systematic errors in the p.e. sequence. For this reason Arp's primary sequence has been remeasured photoelectrically.

The NGC 371 region is very crowded. Some of Arp's stars are only separated by about 10″ arc and thus considerable care has to be taken over the choice of sky background positions. Nights of good seeing alone can be used. The p.e. measures of Arp's sequence have already been discussed at the Prague IAU meeting – the main conclusions were that Arp's sequence does have a systematic error, in that he measured the fainter stars as being too bright. However Arp gives only his adopted (photographically smoothed) measures for some of the faint stars and it is just possible that some part of the error lies in the comparison of p.e. and p.g. measures. This paper presents primarily results of photographic photometry based on plates taken at the Newtonian focus of the Radcliffe 74-in. reflector stopped down to 44 in. (1.12 m) to increase the size of the photometric field.

The NGC 371 region is of interest apart from the cepheids around it, and a more extensive p.e. sequence has been measured in U, B and V down to V = 17.5. At magnitudes fainter than this the crowding problem becomes extremely severe and for a study of the population of the SMC to fainter limiting magnitude a less crowded region would be preferable.

In conjunction with the new p.e. sequence a series of photographic plates has been secured. It is hoped eventually to measure all stars to V = 17.5 in a field of diameter 16′ arc but as this contains many thousands of stars only a very preliminary selection has as yet been made. This consists of photographic measures of Arp's sequence; measures of 3 very small clusters in the region, and measures of all the bright stars down to V = 14.3.

1. Arp Sequence

Arp's sequence and the new p.e. sequence have been measured on all the available photographic plates. The iris curves have been drawn in the usual manner and photographically smoothed magnitudes have been derived for all Arp's sequence

Muller (ed.), The Magellanic Clouds, 79–87.

stars. The changes from the p.e. results are small and random as would be expected.

In addition to the normal photographic plates, a series of plates were taken using an objective grating as described by Wesselink (1964). The grating gives first order images which are 4.85 mag. fainter than the zero order images. It has been shown that the zero order and first order images obey the same colour equation. The main objective of using grating plates, as outlined by Wesselink, is to provide a rapid calibration to faint magnitudes using a p.e. sequence of stars about 5 mag. brighter. In the case discussed here it has been used to give an independent check on the p.e. photometry of the faint stars.

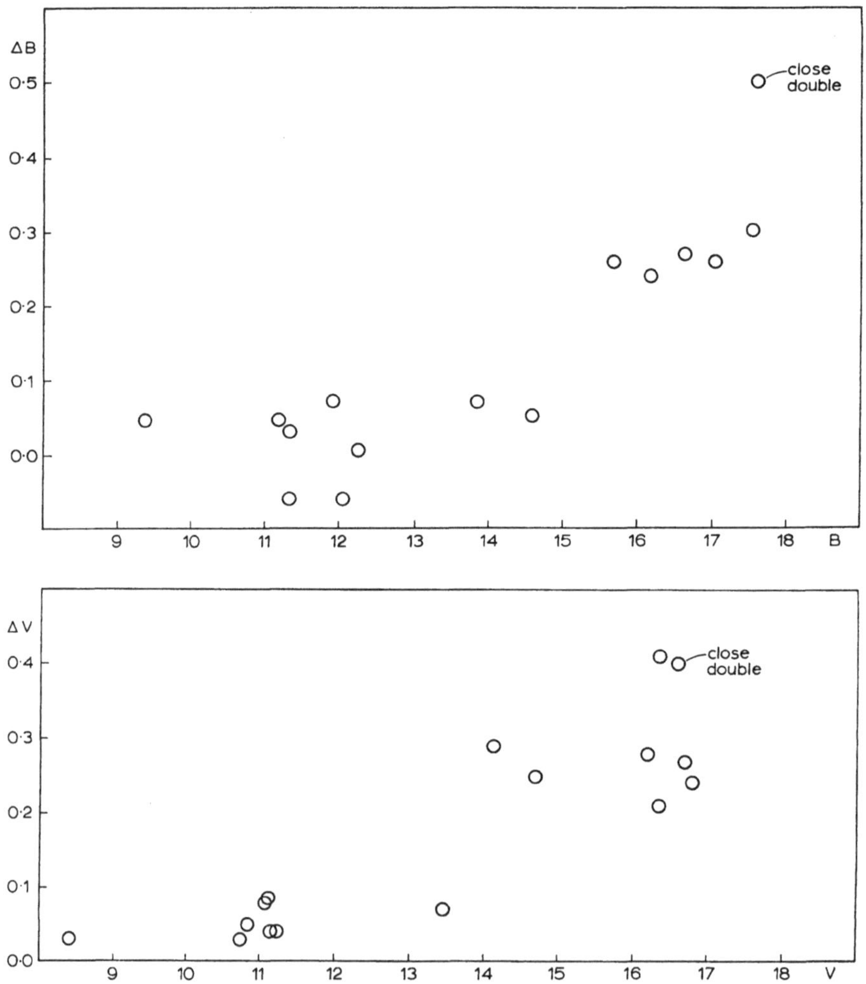

Fig. 1. The differences, ΔV and ΔB, between measures made by Arp and the author, of Arp's p.e. sequence in the NGC 371 region.

TABLE I

Photometry of Arp's NGC 371 sequence

1	2			3			4				5			6	
Arp design	Arp p.e.			Arp adopted			Andrews p.e.				Andrews p.g.[1]			Andrews grating[2]	
	V	B	B−V	V	B	B−V	V	B	B−V	n	V	B	B−V	V	B
H	8.40	9.38	0.98				8.43	9.43	0.995	6					
e	11.38	11.40	0.02				11.35	11.34	−0.01	4					
b	11.22	11.32	0.10				11.26	11.35	0.09	5	11.23	11.34	0.11		
S	10.68	11.16	0.48				10.71	11.20	0.49	6	10.74	11.20	0.46		
U	10.81	12.13	1.32				10.86	12.07	1.21	3					
X	11.00	10.87	−0.13				11.08	10.94	−0.14	4	11.06	10.96	−0.10		
a	11.09	12.25	1.16	11.05	12.27	1.22	11.13	12.27	1.14	6	11.12	12.27	1.15		
b	13.91	15.48	1.57	13.86	15.44	1.58	14.12	15.70	1.58	7	14.15	15.78:	1.63:		15.80
c	13.35	13.82	0.47	13.34	13.80	0.46	13.50	13.97	0.47	7	13.41	13.87	0.46		
d[3]	16.72	16.42	−0.30	16.56	16.38	−0.18	16.80	16.78	−0.02	6	16.79	16.65	−0.14	16.82	16.62
e[4]	15.91	17.26	1.35	15.93	17.25	1.32	16.40	17.38	0.98	6	16.34	17.55	1.21	16.35	17.55
f				14.43	14.56	0.13	14.66	14.58	−0.08	5	14.65	14.61	−0.04		
h[5]				16.46	16.80	0.34	16.53	16.81	0.28	1	16.73	17.06	0.33		17.07
i				16.15	16.01	−0.14	16.32	16.26	−0.06	5	16.36	16.25	−0.11	16.34	16.21
j[6]				16.17	17.07	0.90	16.46	17.01	0.55	5	16.58	17.57:	0.99:	16.48	
k				15.95	15.95	0.00	16.20	16.14	−0.06	5	16.23	16.19	−0.04	16.23	16.23

Notes to Table 1:

[1] The values in column 5 were derived using photoelectric measures of the Arp stars together with other stars chosen to improve and extend the sequence.

[2] The values in column 6 were derived using the $4.^{m}85$ grating and are independent of the measures in column 5.

[3] The corrected Arp p.e. measures are from a private Communication from Arp.

[4] Arp e is only 20″ from Arp a and 15″ from Arp c. The uncertain sky values to be subtracted in the photoelectric observations give large errors especially in the blue.

[5] Arp h was mis-identified by the Author and only one p.e. observation was made of the correct star.

[6] Arp j is a close double, one star being red, the other blue. On most 74″ plates the star is not measurable in the blue.

Table I summarizes the results of both photoelectric and photographic photometry of Arp's sequence.

The photographic magnitudes determined from these two independent methods show no systematic differences and the random errors in both B and V are less than 0.05 mag. for all the stars in Arp's sequence which can be measured.

All the photographic measures have been combined and are shown in Figure 1.

This shows the differences ΔB and ΔV plotted against the B and V magnitudes respectively. The difference is in the sense that Arp measured the faint stars too bright in both B and V. These differences increase to fainter magnitudes but it is difficult, especially for the V magnitudes, to know exactly how to correct the Arp cepheid magnitudes. The sense of the correction is to increase the slope of his period-luminosity relation but it is not certain how much the slope will be increased. Any increase will however bring his relation more into line with those found by others.

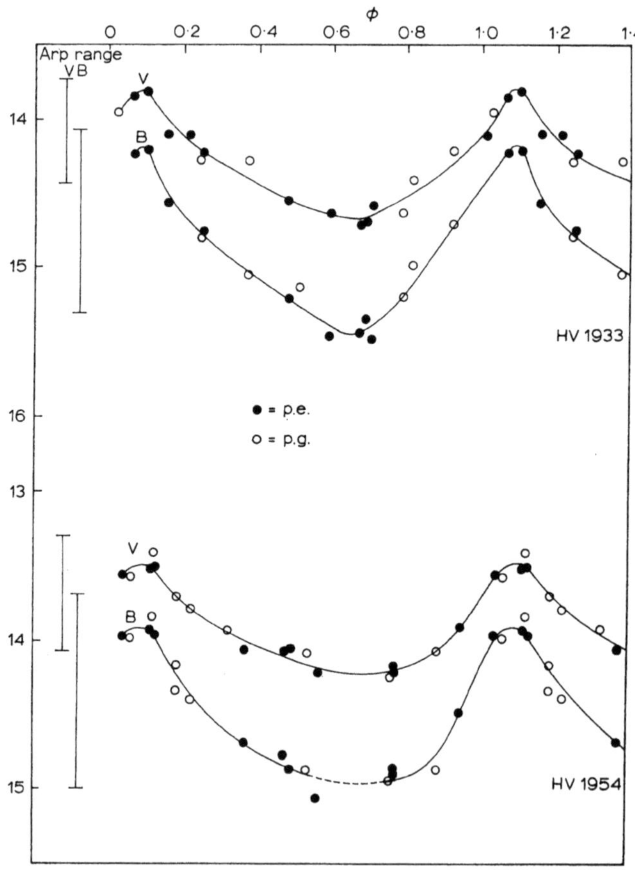

Fig. 2. Combined p.e. and p.g. measures of the SMC cepheids HV 1933 and HV 1954. The magnitude ranges given by Arp are shown as bars.

2. Cepheids

4 Bright cepheids from Arp's list which lie in the NGC 371 region have been observed photoelectrically and photographically and the results are shown in Figures 2 and 3. The ranges given by Arp for these stars are shown and it can be seen that these are brighter than the observed ranges, confirming the above result.

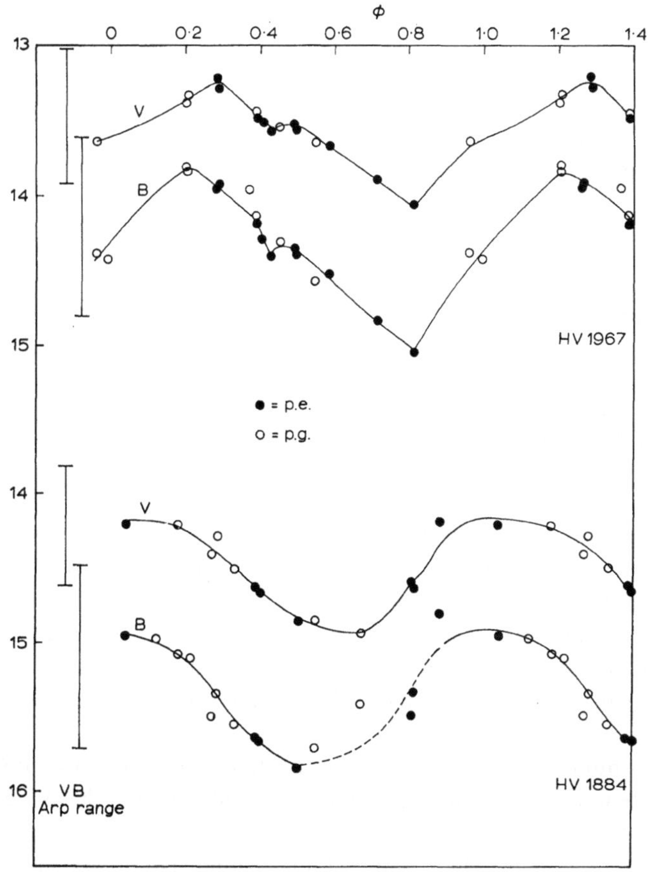

Fig. 3. Combined p.e. and p.g. measures of the SMC cepheids HV 1967 and HV 1884. The magnitude ranges given by Arp are shown as bars.

3. cm Array of Bright Stars

The NGC 371 region has long been known to contain a large number of early type stars, several of the brightest of these have Radcliffe spectra. All the stars inside the photometric field (16′ arc diam) down to V=14.3 have been measured in B and V. The resulting cm array is shown in Figure 4.

This shows the expected group of early-type supergiants which are the top of a near vertical, young-star type of cm array; associated with this is a scatter of stars trailing

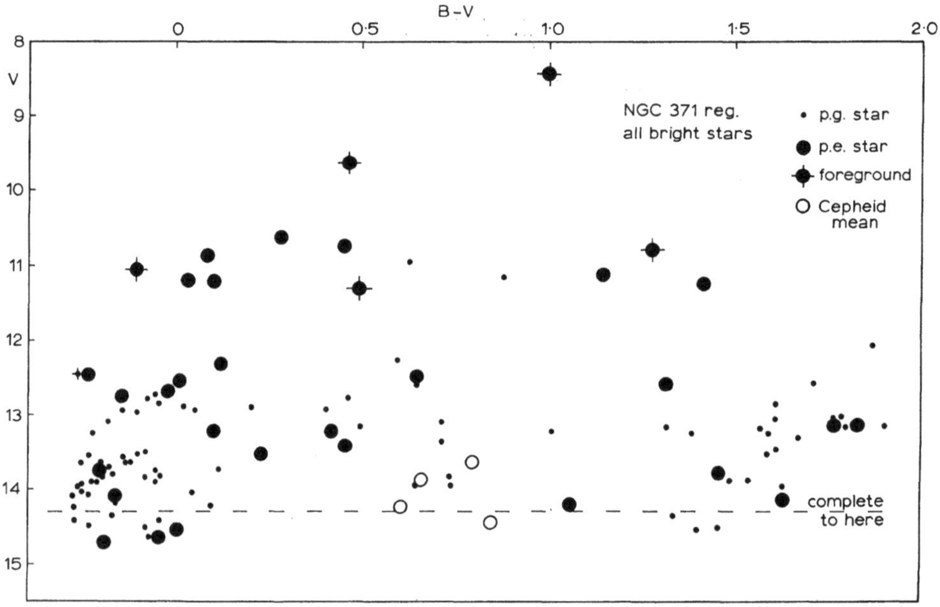

Fig. 4. Colour magnitude array for all stars brighter than V=14.3 inside a 16′ diameter field around NGC 371. Known foreground stars are indicated with crosses. The four open circles are the mean values for the four cepheids in Figures 2 and 3.

off towards the red. There is then a gap in which are the mean positions of the 4 cepheids mentioned above and then another concentration of stars around B−V=1.7.

The reddening for the region has been determined in 3 different ways. The first uses the observed UBV colours for the supergiants, the second the colours of the fainter p.e. standards and the third the photographically determined colours of stars in two clusters which will be described later. All three methods yield values very close to 0.10 for E_{B-V}.

The expected number of foreground stars is about 10 for the area photometered. This agrees quite well with the number found in a similar area in the Wing of the SMC for which the SMC members are all extremely young – the number of non members to V=14.3 in that area is 15. The number of intermediate colour stars in the NGC 371 diagram is 30 and so it is very likely that some at least of these are SMC members.

The similarity of the NGC 371 cm array and the combined arrays of NGC 4755, 3293 and h and χ Per (Feast, 1963) is very striking indicating that this region must be a very populous region of similar age to these groups.

4. cm Array of 3 Small Clusters

These 3 clusters are shown in Figure 5 in relation to NGC 371. They are all small – the

→

Fig. 5. Chart of the NGC 371 region showing the positions of the three small clusters C1, C2 and C3.

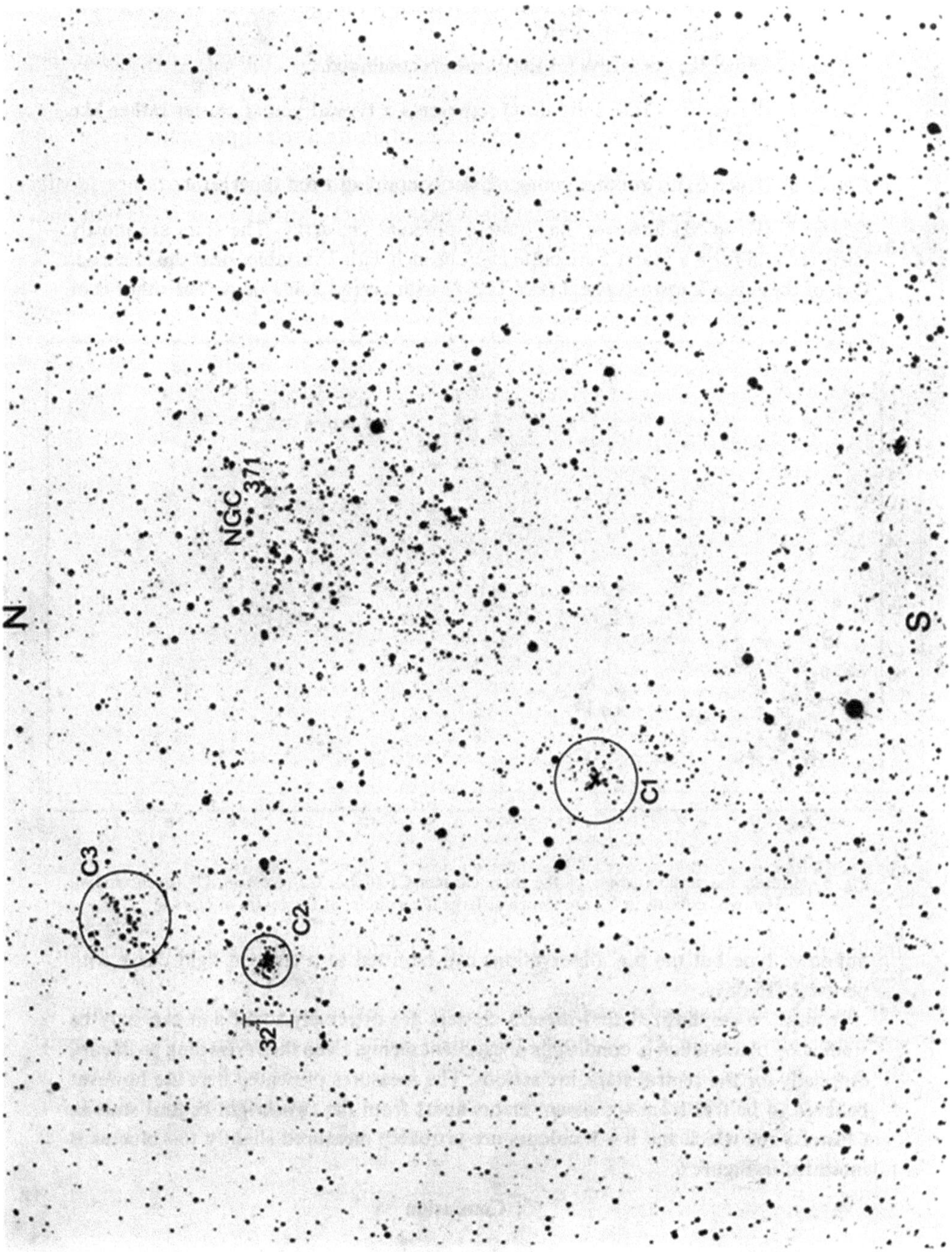
N
S
NGC 371
C1
C2
C3
32

smallest is of about the same apparent diameter as would be expected for the Pleiades if placed at the distance of the SMC.

Figure 6 shows the cm arrays for the 3 clusters combined.

Cluster 1. (Kron 50) – (the solid dots) represents a typical young cluster rather like NGC 2362.

Cluster 3. (Kron 51) is another young cluster containing a red supergiant.

Cluster 2. (Kron 52) however, has quite a different cm array. The stars are mostly fairly red and form a nearly horizontal giant branch with 2 variables near the blue end. One of these is a known cepheid (HV 11201) with period 2.403 days. The other is of

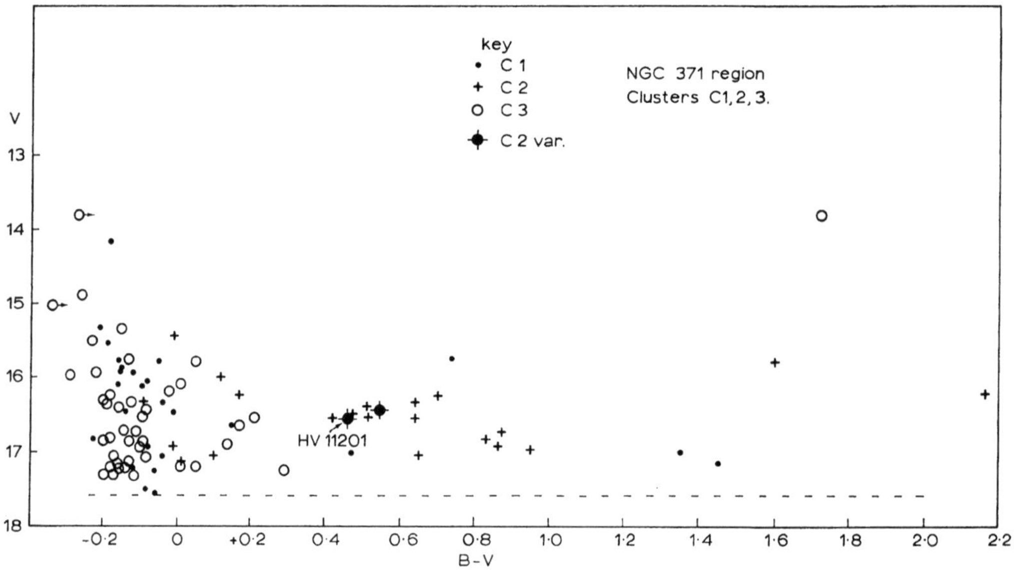

Fig. 6. Colour magnitude arrays of the three clusters C1 (dots), C2 (crosses), C3 (open circles). The two cepheids in C2 are shown as large filled circles at the centre of crosses.

unknown type but the p.g. observations can be fitted to a cepheid light curve with period 2.416 days.

It must be emphasized that these 3 clusters are extremely small and can only be studied on plates taken in conditions of excellent seeing. Even then crowding problems, especially for the central stars, are serious. The measures presented here are however believed to be free from significant errors apart from the two bright central stars in Cluster 3 for which the B – V colours are probably measured slightly too blue as is indicated in Figure 6.

5. Conclusion

The NGC 371 region is dominated by the supergiant stars and these represent a very young population. Cluster 2 indicates that there is also present an older class of stars and it has been seen in the paper by Lloyd Evans (see page 74) that there are also indications of old giant stars in this region. The presence of what appear to be normal globular clusters in the SMC leads one to the conclusion that there is a range in both age and in chemical composition comparable with those in the galaxy.

References

Arp, H. C.: 1960, *Astron. J.* **65**, 404.
Feast, M. W.: 1963, *Monthly Notices Roy. Astron. Soc.* **126**, 11.
Feast, M. W., Thackeray, A. D., and Wesselink, A. J.: 1960, *Monthly Notices Roy. Astron. Soc.* **121**, 345.
Gascoigne, S. C. B. and Kron, G. E.: 1965, *Monthly Notices Roy. Astron. Soc.* **130**, 333.
Wesselink, A. J.: 1964, *Appl. Opt.* **3**, 889.
Woolley, R. v. d. R. *et al.*: 1962, *Roy. Observ. Bull.* **58**.

A SPECTROSCOPIC STUDY OF MAGELLANIC CLOUD GLOBULAR-TYPE CLUSTERS

P. J. ANDREWS and T. LLOYD EVANS
Radcliffe Observatory, Pretoria, South Africa

Several investigations of the integrated spectra of globular clusters in the Galaxy have been carried out. The spectral types assigned on the basis of low dispersion spectra are correlated with other parameters, which in turn are believed to reflect differences in metal abundance, and hence provide a convenient method of differentiating between clusters of high or low metal abundance. A further use is the study of the kinematics of the globular cluster system. The present paper deals with the spectral classification of red globular clusters in the Large and Small Magellanic Clouds.

Spectra of 9 LMC and 4 SMC red globular clusters have been secured at a dispersion of 312 A/mm on 103aO film using the spectrograph described by Evans (1963) at the Newtonian focus of the 1.88 m reflector. Spectra of 16 globular clusters in our own Galaxy, most of which have spectral types assigned by Kinman (1959), were observed for comparison.

Spectral types were assigned independently by the two authors and an average taken. Separate estimates of spectral type from the ratio of the G band to Hδ Sp(G), (Hγ falls too close to the mercury line at λ 4358) and, Sp(H), the hydrogen line strength relative to the H and K lines of Ca II were made. The good ultraviolet transmission of the spectrograph permits the use of the near UV as far as 3700–3800 Å on spectra of average density. One might expect the hydrogen line spectral types to be more affected by the degree of compositeness (related to blue horizontal branch strength in particular) than were those of Kinman who used a spectrograph with poor UV transmission. However, little difficulty was experienced in ordering the spectra of galactic globular clusters in the same way as Kinman, except for two cases. Our final types should be very close to his system. Agreement between the two authors was good, the largest difference being 2 subclasses.

The spectra of Kron 3 in the SMC and Hodge 11 in the LMC show H 9 and H 10 unusually weak by comparison with H 8, these two clusters are discussed separately below.

The remaining clusters allow the following conclusions to be drawn:

(1) The $(B-V)$ colours (van den Bergh, 1968) show a dependence on spectral type similar to that for galactic globular clusters (van den Bergh, 1967). The reddening shows a total range of only about $0^m.2$ and has average values for the two clouds of

$$\text{SMC } E_{B-V} = 0^m.14 \text{ (3 clusters)}$$
$$\text{LMC } E_{B-V} = 0^m.09 \text{ (7 clusters)}$$

in good agreement with the values determined from UBV observations of supergiants.

Muller (ed.), The Magellanic Clouds, 88–91.

(2) Van den Bergh's metallicity index Q shows no definite correlation with G band or Hδ spectral type, in accordance with the lack of correlation of Q with $B-V$.

(3) All the clusters observed so far have spectral types indicative of moderate metal deficiency, with none so early or so late as the extreme examples in our own Galaxy. This could result from the smallness of the sample or from the selection by apparent brightness.

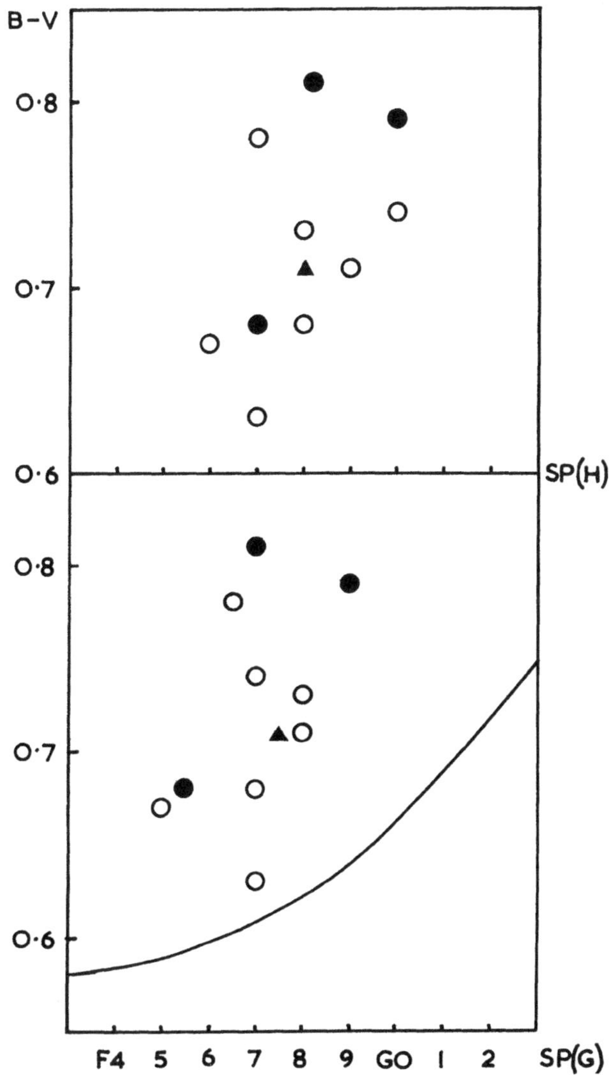

Fig. 1. $B-V$ plotted against hydrogen and G-band spectral types, respectively. Open symbols denote LMC clusters and closed symbols SMC clusters, and Kron 3 is represented by a triangle. The solid line is van den Bergh's intrinsic relation for galactic globular clusters.

(4) The dependence of the spectral type difference, Sp(H) − Sp(G), on Sp(G) is indistinguishable from that found for galactic globular clusters so that the degree of compositeness of the spectra (and hence horizontal branch strength) must be fairly similar for given Sp(G).

Gascoigne (1966) found that NGC 121, 1466 and 2257 have colour magnitude diagrams similar to those of globular clusters in the Galaxy. The occurrence of RR Lyrae variables in these as well as in NGC 1978 (Thackeray, 1960) strengthens the similarity. The spectral types assigned to these clusters are appropriate to their colour magnitude diagrams.

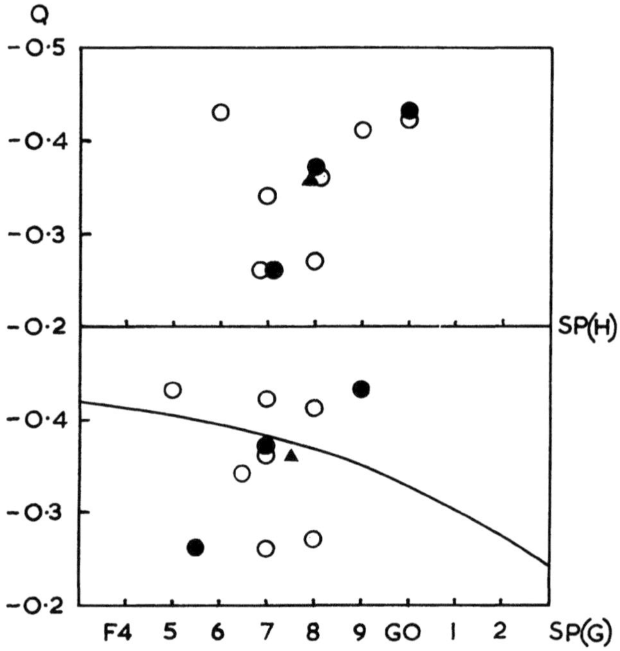

Fig. 2. Van den Bergh's Q index plotted against hydrogen and G-band spectral types, respectively. Open symbols denote LMC clusters and closed symbols SMC clusters. Kron 3 is represented by a triangle. The solid line is van den Bergh's intrinsic relation for galactic globular clusters.

Gascoigne regarded the colour magnitude diagram of NGC 1783 as being peculiar but our spectra show no obvious peculiarity, we note that it is bluer for its spectral type than any of the other clusters for which we have data.

The two clusters Kron 3 and Hodge 11 were both noted by Gascoigne to be peculiar. Our spectra show very strong H and K, a moderately strong G-band, moderately strong H 8 and Hδ, but H 9 and H 10 are very weak. There is a possibility that these effects may be caused by night sky (including terrestrial) contamination on these long, rather weak exposures.

Gascoigne regarded Kron 3 as belonging to an 'SMC group' possibly similar to

TABLE I

Cluster		Sp(G)	S_p(H)	$(B-V)$	$(B-V)_{STD}$	E_{B-V}	Q
NGC	121	F 9	G 0	0.79	0.64	0.15	−0.43
	416	F 7	F 8	0.81	0.61	0.20	−0.37
	419	F 5–6	F 7	0.68	0.59	0.09	−0.26
KRON	3	F 7–8	F 8 p	0.71	(0.62)	(0.09:)	−0.36
					Mean = *0.14 (3 clusters)*		
NGC	1466	F 7	F 8	0.68	0.61	0.07	−0.36
	1754	F 5	F 7		0.59		
	1783	F 7	F 7	0.63	0.61	0.02	−0.26
	1786	F 5	F 6	(0.67)	0.59	(0.08)	(−0.43)
	1806	F 8	F 8	0.73	0.63	0.10	−0.27
	1835	F 6–8	G 0	0.74	0.62	0.12	−0.42
	1978	F 6–7	F 7	0.78	0.61	0.17	−0.34
	2210	F 8	F 9	0.71	0.63	0.08	−0.41
Hodge	11	F 6–7	F 8 p		(0.61)		
					Mean = *0.09 (7 clusters)*		

open clusters of intermediate age. Walker (1970) has however derived a colour magnitude diagram similar to metal rich clusters such as 47 Tuc, but showing other peculiarities including a very red tip to the giant branch which would not be expected if the cluster were like 47 Tuc. Our spectra are significantly different from those of 47 Tuc.

NGC 419 which Gascoigne assigns to his SMC group on the basis of Arp's photometry has a typical halo type spectrum quite unlike that of Kron 3.

Hodge 11 is an example of Gascoigne's 'LMC group' with a very poorly defined giant branch; as our spectrum will be dominated by subgiants and horizontal branch it is difficult to correlate the various peculiarities.

References

Arp, H. C.: 1958, *Astron. J.* **63**, 273.
Bergh, S. van den: 1967, *Astron. J.* **72**, 70.
Bergh, S. van den and Hagen, G. L.: 1968, *Astron. J.* **73**, 569.
Evans, D. S.: 1963, *Monthly Notices Astron. Soc. South Africa* **22**, 140.
Gascoigne, S. C. B.: 1966, *Monthly Notices Roy. Astron. Soc.* **134**, 59.
Kinman, T. D.: 1959, *Monthly Notices Roy. Astron. Soc.* **119**, 538.
Thackeray, A. D.: 1960, *Astron. J.* **64**, 437.
Walker, M. F.: 1970, *Astrophys. J.* **161**, 835.

L'ÉMISSION Hα DANS LE GRAND NUAGE DE MAGELLAN. PROGRAMMES EN DÉVELOPPEMENT À CÓRDOBA*

GUSTAVO J. CARRANZA
Observatorio de Córdoba – C.N.I.C.T.

Les programmes d'observation qu'on développe actuellement à Córdoba sur les Nuages de Magellan employent plusieurs instruments dont voici la liste:

(1) Lunette de 8 cm, $F/5$
(2) Lunette de 20 cm, $F/5$
(3) Télescope de 154 cm, $F/5$.

Tous ces instruments peuvent travailler soit avec un réducteur focal WRAY $F/1.25$ (clichés monochromatiques), soit avec un interférometre de Pérot-Fabry.**

Le but général de ces programmes est d'étudier les propriétés du champ d'émission Hα général de ces galaxies. Plus précisément, on cherche à établir:

(1) une carte de l'émission Hα générale au moyen de filtres interférentiels étroits (HW 8–10 Å);

(2) leur champ de vitesses, avec toutes les avantages de l'interférometrie de Pérot-Fabry (principalement dispersion et simultanéité sur un champ étendu).

L'emploi d'une batterie d'instruments à une game de résolutions aussi variée que celle qu'on a décrite permet de ne pas perdre de vue le caractère de galaxies des Nuages de Magellan, ainsi que d'étudier les propriétés individuelles de leurs différentes régions H II.

A ce stade, l'effort est porté sur le probleme de la détection de l'émission globale; on espère prochainnement développer les observations cinématiques. Jusqu'a maintenant les travaux d'observation ont été compliqués par les délais imposés par la mise en opération à Córdoba des appareils, et principalement par le mauvais temps.

Nous pouvons néanmoins énoncer déjà quelques résultats intéressants concernant l'émission Hα du Grand Nuage.

Les travaux faits par Johnson [1] (basse résolution et basse sélectivité) et par Cruvellier [2] (haute sélectivité mais aussi résolution un peu trop grande) par exemple, ont mis en évidence l'importance de l'émission Hα du grand complexe de 30 Doradus. Cruvellier a même trouvé des grandes extensions en émission dans le voisinage de cette nébuleuse.

Les observations que nous avons faites avec l'instrument de 8 cm nous ont permis

* Avec le concours du Consejo Nacional de Investigaciones Científicas y Técnicas, Buenos Aires.
** Équipement réalizé par le Service d'Interférometrie de l'Observatoire de Marseille et le Laboratoire d'Astronomie Spatiale du CNRS à Marseille, France.

Muller (ed.), The Magellanic Clouds, 92–94.

de déterminer que cette émission générale (Figure 1*) s'étale partout dans une région d'environ 1° × 2° entre 30 Doradus et S Doradus, un peu au Nord du corps principal du Grand Nuage (basse).

Cette faible émission, loin d'être uniforme, possède une structure filamentaire (semblable a des lignes de courant), les filaments se ramifiant vers les régions H II brillantes (30 Dor, S Dor, Henize 144) a partir d'une région placée à l'extrémité SE de la barre, aux coordonnées 5h 24m et −70°.

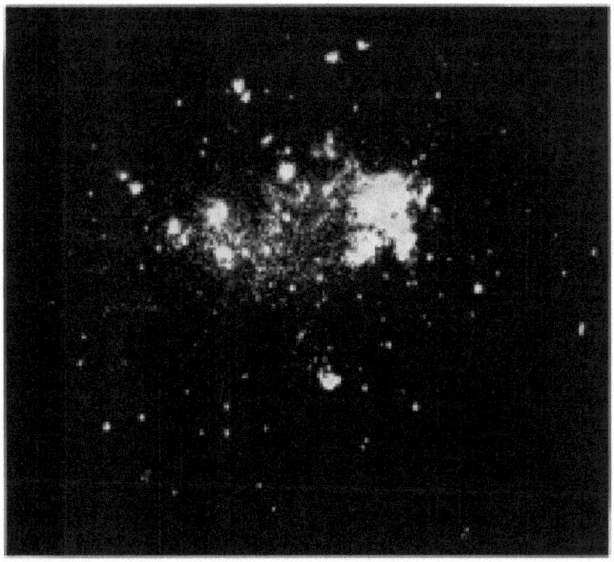

Fig. 1.

Les tres longes poses que demanderait l'obtention d'interférogrammes de cette émission générale (decelée sur plaque IaE préexposée et avec 4h 30m de pose) nous obligent d'attendre les prochaines saisons d'observation pour l'étudier interférometriquement et ainsi d'avoir une idée plus claire de ses rapports avec la structure déjà connue du Grand Nuage et avec l'émission générale qu'on a trouvée dans d'autres galaxies (M 33[3] et M 83[4] par exemple).

Concernant l'étude avec le réflecteur de 154 cm des régions H II individuelles, on possède déjà un certain nombre des résultats (Carranza et Monnet, à publier) sur quelques nébuleuses (Henize 11 B, C, D; 44 C, D, E, F, H, I; 51 A, C, D, E; 59 A, B; 159; 160 du Grand Nuage, Henize 36; 37 du Petit Nuage).

La comparaison de nos vitesses avec celles mesurées par Feast [5] et par McGee et Milton [6] montre un tres bon accord, les écarts systematiques étant de +1.7 km/sec et de +4.5 km/sec respectivement. Cela correspond peut-être, a un effect d'expansion des régions H II.

* Cette illustration a été faite sur film lithografique à très grand contraste.

References

[1] Johnson, H. M.: 1959, *Publ. Astron. Soc. Pacific* **71**, 301.
[2] Cruvellier, P.: 1968, *Ann. Astrophys.* **30**, 1059.
[3] Carranza, G., Courtes, G., Georgelin, Y., Monnet, G., et Pourcelot, A.: 1968, *Ann. Astrophys.* **31**, 63.
[4] Carranza, G.: 1968, XIV Boletín de la Asociación Argentina de Astronomía.
[5] Feast, M. W.: 1964, *Monthly Notices Roy. Astron. Soc.* **127**, 15.
[6] McGee, R. X. et Milton, J. A.: 1966, *Australian J. Phys.* **19**, 343; et 1966, *Australian J. Phys. Suppl.* **2**.

OPTICAL ASPECTS OF X-RAY SOURCES IN THE LARGE MAGELLANIC CLOUD

HUGH M. JOHNSON
Lockheed Missiles and Space Co., Palo Alto, Calif., U.S.A.

The detection of X-rays from the Large Cloud by Mark *et al.* (1969) leads to the question of optical aspects on the assumption that a power-law spectrum joins the radiofrequency and X-ray data. The integrated nonthermal radiofrequency spectrum (Mathewson and Healey, 1964) may extrapolate smoothly to the X-ray datum in the same way that the corresponding data of Vir A, Cas A, and Tau A do (Figure 1). But the integrated nonthermal radiofrequency spectrum of the Small Cloud (Mathewson and Healey, 1964) lies one order of magnitude lower than that of the Large Cloud in flux density, so that a corresponding X-ray source would not have detected by Mark *et al.* in agreement with their negative observation of it.

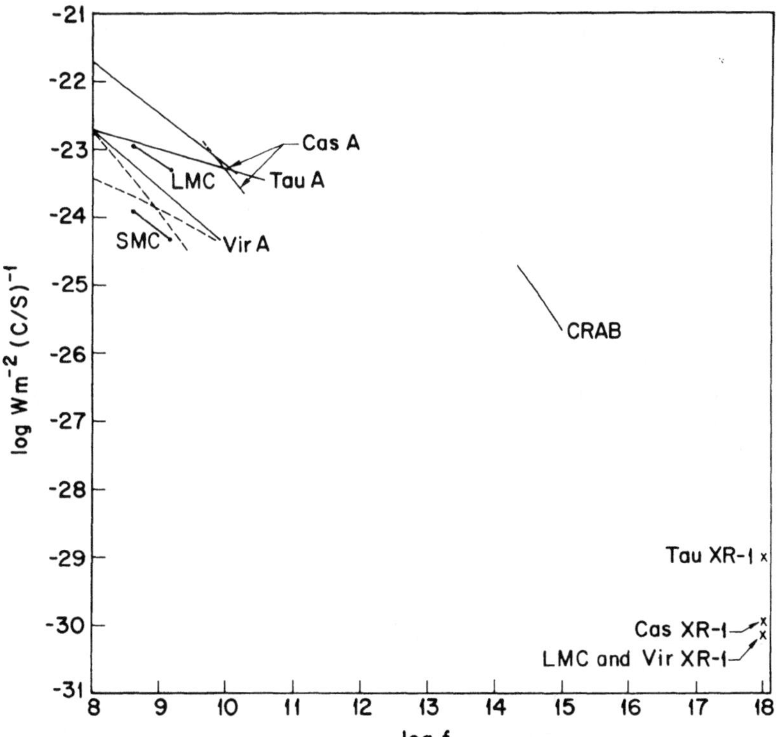

Fig. 1. The continuous spectra of several radio/X-ray sources. The optical continuum of Vir A, not shown here, is discussed by Felten (1968). Radio spectra are compiled from many sources, and X-ray data except of the LMC are based on Friedman *et al.* (1967).

Muller (ed.), The Magellanic Clouds, 95–97.

The optical continuum of the Crab nebula lies on the spectrum interpolated between the flux densities of Tau A and Tau XR-1 (Figure 1), and the jet of M 87 is probably part of the spectrum likewise interpolated between the flux densities of Vir A and Vir XR-1. The optical continuum may also be present in Cas A/Cas XR-1 except much weakened by interstellar extinction (Johnson 1968, and unpublished observations). Is a nonthermal optical continuum visible in the Large Cloud?

Extrapolated to visual frequencies the flux density of the radiofrequency spectrum is 2.3×10^{-27} Wm^{-2} Hz^{-1}. The formula $\log F_\nu = -0.4\,V - 22.42$, derived from Oke (1966), converts flux density in these units to visual magnitude V. Then $V = 10.5$ is the integrated visual magnitude of the continuous nonthermal spectrum assumed in the Large Cloud. This is 9.8 mag fainter than the observed integrated visual magnitude of the Large Cloud. The nonthermal radiofrequency emission (Mathewson and Healey, 1964) and X-radiation (Mark *et al.*, 1969) is distributed on the sky roughly like the optical light of the whole Large Cloud, but 30 Dor and three other optical objects have been identified as localized nonthermal sources (Westerlund and Mathewson, 1966). Table I gives some of their parameters. We conclude that objects like N63A and N132D might be visible in the nonthermal continuum. N36A and N49 are plainly visible on the charts of the *Uppsala-Mount Stromlo Atlas of the Magellanic Clouds* (Gascoigne and Westerlund, 1961) and on the charts of *The Large Magellanic Cloud* (Hodge and Wright, 1967). N132D is dubious because of crowding by stars.

TABLE I

Local nonthermal sources in the Large Cloud

Object	Dimension	Spectral index	Observed F_ν(408 MHz)	Extrapolated F_ν(visual)	Extrapolated V mag	Mean surface brightness V mag/□″
			units of 10^{-26} Wm^{-2} Hz^{-1}			
30 Dor	24′ FWHM at 408 MHz	−0.6	(210)	4.4×10^{-2}	12.3	27.8
N49	67″ FW at Hα	−1.0	9	6.7×10^{-6}	21.8	30.6
N63A	27″ FW at Hα	−0.5	3	2.6×10^{-3}	15.4	22.3
N132D	22″ FW at Hα	−0.5	(7)	6.0×10^{-3}	14.5	20.9

The plates for these charts were exposed on 103a-D emulsion behind GG-11 or GG-14 filters, hence cover the approximate range $\lambda\lambda$ 4900–5800 Å. They may, of course, leak some emission lines. Little if anything has been done to look for continuous-spectrum nebulae in the Clouds using passbands which would strictly exclude emission lines. Westerlund and Mathewson (1966) found no trace of continuum in their spectrograms. If the nonthermal continuum is *not* visible in objects such as N63A and N132D, we may conclude that such objects have little chance to contribute to the X-ray emission of the Large Cloud. We also note that 30 Dor has been likened to the nucleus of a galaxy on the basis of optical and radio data (Johnson, 1959) and again on the basis of further radio data (Mathewson and Healey, 1964). However, the nucleus of the Galaxy has yet to be detected as an X-ray source.

Acknowledgement

This work has been done under the Lockheed Independent Research Program and under Office of Naval Research Contract N00014-69-C-0147.

References

Felten, J. E.: 1968, *Astrophys. J.* **151**, 861.
Friedman, H., Byram, E. T., and Chubb, T. A.: 1967, *Science* **156**, 374.
Gascoigne, S. C. B. and Westerlund, B. E.: 1961, *Uppsala-Mount Stromlo Atlas of the Magellanic Clouds*, Australian National University, Canberra, A.C.T.
Hodge, P. W. and Wright, F. W.: 1967, *The Large Magellanic Cloud*, Smithsonian Press, Washington, D.C.
Johnson, H. M.: 1959, *Publ. Astron. Soc. Pacific* **71**, 301 and 425.
Johnson, H. M.: 1968, *Can. J. Phys.* **46**, S481.
Mark, H., Price, R., Rodrigues, R., Seward, F. D., and Swift, C. D.: 1969, *Astrophys. J. (Letters)* **155**, L143.
Mathewson, D. S. and Healey, J. R.: 1964, in *The Galaxy and the Magellanic Clouds* (ed. by F. J. Kerr and A. W. Rodgers), Australian Acad. of Sci., Canberra, A.C.T., p. 245.
Oke, J B.: 1966, *Astrophys. J.* **145**, 668.
Westerlund, B. E. and Mathewson D. S.: 1966, *Monthly Notices Roy. Astron. Soc.* **131**, 371.

RADIO CONTINUUM OBSERVATIONS OF THE MAGELLANIC CLOUDS

D. S. MATHEWSON
Mount Stromlo and Siding Spring Observatories Research School of Physical Sciences
The Australian National University

1. Stars, Gas, Magnetic Fields and Cosmic Rays

A. NON-THERMAL EMISSION

A galaxy is made up of a number of different ingredients: stars, gas, magnetic fields and cosmic rays. This talk will be about observations of the radio continuum emission from the Magellanic Clouds which is produced by the synchrotron mechanism when the last two ingredients are mixed together. Measurements of the intensity, spectra, and polarization of this radio emission can tell us something about these two invisible constituents which we know from radio observations of our Galaxy are found both in extended regions (spiral arms) and in discrete sources (supernova remnants).

It is now widely recognized that a knowledge of the structure and strength of the magnetic fields is essential before problems related to galactic structure, evolution and dynamics can be solved. Particularly as Verschuur (1969) has recently measured magnetic field strengths as high as 2×10^{-5}G in several regions of our Galaxy. Therefore it is quite likely that the magnetic field is dynamically important in individual spiral arms and perhaps even dynamically important on the scale of a galaxy as a whole. Also, the energy density of cosmic rays and their pressure in a number of situations are believed to be a major factor in energetics and dynamics.

B. THERMAL EMISSION

This talk will also concern the radio observations of the thermal emission from H II regions, i.e., regions where the first two ingredients, young stars and gas, are mixed together. These are more closely related to optical observations and their main importance is that they are not hampered by obscuring dust clouds. In fact, there are several thermal radio sources, 50 to 100 pc in diameter, in the Clouds that have no optical identification.

2. Foreground Galactic Radio Emission

It is important to consider the general structure of the foreground galactic radio emission in the direction of the Magellanic Clouds as this determines how much information can be obtained about their intrinsic emission.

The survey at 85 Mc/sec by Yates (1968) shows that the Large Magellanic Cloud (LMC) at $l^{II}=278°$, $b^{II}=-32°$ lies near one of the deepest minima in the radio sky. The rado isophotes are fairly regular in this region which allows an accurate estimate of the radio emission from the LMC. The reason for this deep minimum in the galactic

Muller (ed.), The Magellanic Clouds, 98–108. *All Rights Reserved.*

radio emission is that the line-of-sight component of the local magnetic field is large in this direction and therefore the synchrotron emission is weak (Mathewson and Nicholls, 1968). Thus the measured polarization of the non-thermal radio emission from the LMC is very weak due to large depolarization effects by differential Faraday rotation over the aerial beam.

The Small Magellanic Cloud (SMC) at $l^{II} \approx 302°$, $b^{II} = -44°$ is in a direction where the line-of-sight component of the galactic magnetic field is smaller but, because of this, it lies in a region of strong galactic polarization (Mathewson and Milne, 1965) which makes the measurement of the intrinsic polarization uncertain. In addition, the SMC lies in the direction of a strong galactic radio spur and the steep gradients make it difficult to separate the SMC and the galactic radio emission.

3. Radio Observations of the Magellanic Clouds

Table I gives details of the radio observations of the Large and Small Magellanic Clouds. Surveys at 136 Mc/sec, 408 Mc/sec, 1410 Mc/sec, 2650 Mc/sec and 5009 Mc/sec have been made with the Parkes 210-ft reflector. Surveys at 19.7 Mc/sec, 85.5 Mc/sec and 408 Mc/sec have been made with Mills Cross-type arrays at Fleurs and Hoskinstown.

TABLE I
Radio continuum observations

Frequency (Mc/sec)	Aerial beamwidth (min. of arc)	Radio telescope	Observers
19.7	84	Cross Array, Fleurs	Shain (1959)
85.5	50 × 65	Cross Array, Fleurs	Mills (1959)
136	144	210-ft. Reflector, Parkes	Mathewson and Healey (1964a)
408	48	210-ft. Reflector, Parkes	Mathewson and Healey (1964a)
1410	14	210-ft. Reflector, Parkes	Mathewson and Healey (1964a)
2650	7.4	210-ft. Reflector, Parkes	Broten (1965)
5000	4.0	210-ft. Reflector, Parkes	McGee (in process)
408	2.8 × 3.5	Cross Array, Hoskinstown	Mills and Aller (in process) Le Marne (1968)

4. Radio Structure of the Large Magellanic Cloud

The radio structure of the LMC may be conveniently divided into five components: 30 Doradus, the central 'axial-bar', the disk, the discrete sources (H II regions and supernova remnants), and the outer arms.

A. 30 DORADUS

Mathewson and Healey (1964a) proposed a model consisting of a strong central thermal source of 4′ diameter coincident with N157A (Henize, 1956) surrounded by a non-thermal source of 24′ diameter. This in turn is surrounded by an extensive H II region about 45′ between half-power points. Le Marne (1968) using the high resolution of the Mills Cross at 408 Mc/sec has confirmed and refined this model. The non-thermal component is centred slightly to the West of N157A and is $26' \times 13'$ at half-power points. A non-thermal source is found at the position of N157B.

At high frequencies, 30 Doradus dominates the radio picture of the LMC just as the galactic centre dominates the radio picture of our Galaxy. Structurally they are very similar. They each have thermal and non-thermal components present of fairly similar dimensions and emissivities (Table II).

This comparison would show them to be even more similar if a measurement of the spectral index of the central core of 30 Doradus showed that, like Sagittarius A, it was nonthermal at frequencies greater than 3000 Mc/sec. At frequencies lower than 3000 Mc/sec they both have 'flat' spectral indices. In this respect, it is interesting to note that, like Sagittarius A, no hydrogen recombination lines have been detected from the core of 30 Doradus even though the line has been detected in galactic H II regions which give the same continuum temperature (McGee and Gardner, 1968).

From this radio evidence it is difficult to avoid the conclusion that 30 Doradus is the nucleus of the LMC. This is supported by the H-line measurements of McGee and Milton (1966) which show that one, if not two spiral arms, originate from the region of 30 Doradus.

B. THE CENTRAL AXIAL-BAR

This non-thermal component shows clearly on the scans at 136 Mc/sec and 408 Mc/sec (Mathewson and Healey, 1964a). It has a width at half-power of $3° \times 2°$ and appears to be associated with the axial bar. The ridge line of maximum emission runs along the northern edge of the bar. The integrated radio emission at 408 Mc/sec from this component is 160 flux-units which is about 15% of the total radio emission from the LMC. (A flux-unit is 10^{-26} $Wm^{-2}(c/sec)^{-1}$.) The spectral index of the emission is -0.6.

C. THE DISK COMPONENT

This is a very extended non-thermal component roughly the size of the visible galaxy. It has an approximate gaussian brightness distribution and a width at half-power points of $5° \times 7°$. It is responsible for about two-thirds of the total radio emission at 408 Mc/sec with an integrated flux density of 700 flux-units. The spectral index of the radio emission is -0.6. It has approximately one-third of the emissivity of the galactic disk component. It is interesting that this component does not appear to be broken up into spiral arm features.

D. THE DISCRETE SOURCES

The 1410 Mc/sec isophotes (Mathewson and Healey, 1964a) and the 2650 Mc/sec isophotes (Broten, 1965) show a number of discrete sources most of which are identified with emission nebulae in the catalog of Henize (1956). In most cases, reasonable agreement is obtained between the observed and the predicted radio flux densities

TABLE II

Comparison of the galactic center and 30 Doradus

Galactic center components	Size (parsecs)	Spectra	Intensity at distance of LMC[a]	30 Doradus components	Size (parsecs)	Spectra	Intensity[a]
SGR A	10	0.2 $f < 3000$ Mc/sec -0.7 $f > 3000$ Mc/sec	12 f.u. (3000 Mc/sec)	N 157A (central source) N 157B	60 16	0.2 $f < 3000$ Mc/sec ? $f > 3000$ Mc/sec non-thl.	40 f.u. (3000 Mc/sec) 4 f.u. (408 Mc/sec)
Extended thermal components	170×70	th.	24 f.u. (3000 Mc/sec)	Outlying regions of 30 Doradus	Mostly within		
Thermal discrete sources	Within 200 pc of center	th.	Total 16 f.u. (3000 Mc/sec)		300 pc of center	th.	≈ 20 f.u. (3000 Mc/sec)
Extended non-thermal component	500×170	non-th.	330 K (408 Mc/sec)	Extended non-thermal component	400×200	-0.6	350 K (408 Mc/sec)

[a] f.u. = flux units = 1×10^{-26} Wm^{-2} (c/sec)$^{-1}$

calculated from the Hβ (Dickel *et al.*, 1964) and Hα flux densities (Doherty *et al.*, 1956).

Mr. N. W. Broten of National Research Council, Ottawa, has kindly sent me his 2650 Mc/sec isophotes of the LMC and SMC. His maps were convolved with a 12′ of arc gaussian function to make their resolution equal to that of the 1410 Mc/sec map of Mathewson and Healey (1964a). Peak brightness temperatures of the sources were then compared directly to obtain the spectral index of the radio emission. Most sources had spectral indices characteristic of an optically thin thermal source although the known supernova remnants N49, N63A and N132D had non-thermal spectral indices (Mathewson and Healey, 1964a; Westerlund and Mathewson, 1966).

Poveda and Woltjer (1968) have suggested a relationship between the surface brightness and radius of a supernova remnant. Milne (1969) with considerably more data has derived a more accurate relationship which gives the distance to the LMC as 55 kpc using the data for N63A and N132D. This is in good agreement with other methods used for the distance measurement. However, N49, if it is a member of the LMC at a distance of 55 kpc, does not obey the relation. In fact, its surface brightness is about 20 times too large or it is at a distance of 30 kpc.

Westerlund and Mathewson (1966) found for N63A, N132D, Cas A and the Cygnus Loop that the magnetic field strength (calculated assuming equipartition between the high energy particles and magnetic field) was inversely proportional to the square of the radius. This is in agreement with the prediction of Shklovsky (1960) based on the argument that the general structure of the magnetic field inside the supernova remnant remains the same during the expansion, i.e., the magnetic flux is constant and the magnetic field inversely proportional to the square of the radius. (This result does not support the suggestion by van der Laan (1962) that the radio emission from supernova remnants arises from the region of compressed interstellar magnetic field outside the expanding shell.) However for N49, it was found that, for its size, it had a magnetic field 5 times stronger than the other Magellanic and galactic supernova remnants. This could explain its surface brightness being 20 times greater than required by Milne's relation which the other supernova remnants satisfy. These results reinforce the suggestion by Westerlund and Mathewson (1966) that N49 may be the remnant of an O star supernova whereas the others are from the more common B stars.

Table III lists interesting radio sources in the main body and outskirts of the LMC which would be worth further investigation. These are mainly extended sources with thermal type spectra that lie in the region of the LMC and have no optical identification. A concentration of H I is present for many of them which increases the probability that an emission nebula exists at each position (McGee and Milton, 1966). N94 may be a supernova remnant as the spectral index of the integrated emission at 1410 Mc/sec and 408 Mc/sec from the complex of emission nebulae around this position is non-thermal. The high resolution observations of Mills and Aller (1969) at 408 Mc/sec are needed to isolate the non-thermal component.

TABLE III

Interesting radio sources in the region of the Large Magellanic Cloud

Source number	Position (1975) R.A. h m	Dec ° ′	Flux density, 20 cm $\times 10^{-26}$ Wm^{-2} (c/sec)$^{-1}$	Size (min. of arc)	Spectra	Remarks
1	05 31.8	69 04	0.6	<3	th.	
2	05 47.2	69 42	1.0	4	th.	HI concentration
3	05 51.0	68 23	0.7	<3	th.	HI concentration
4	04 49.7	70 52	1.0	>6	th.	HI concentration
5	04 55.7	69 28	(complex)		non-th.	N94, SN remnant?
6	05 15.7	67 21	1.0	3	th.	HI concentration
7	04 36.4	74 51	1.4	11	th.	southern extension, HI present
8	04 58.2	75 05	1.0	6	th.	southern extension, HI present
9	04 54.8	64 48	1.0	<6	th.	HI concentration
10	05 08.0	62 41	1.6	8	th.	
11	06 08.7	65 49	1.1		th.	

E. OUTLYING ARMS

Spurs of emission at 1410 Mc/sec and 408 Mc/sec (Mathewson and Healey, 1964a) extend to the north and south-west from the main body of the LMC. These do not fit in with the general run of the foreground galactic isophotes and it is suggested that they indicate outlying arms of the LMC. In addition, they contain some extended thermal radio sources which at this high latitude strongly suggest that these features are associated with the LMC.

In the north, there are two spurs. The first starts at 05^h30^m, $-65°$ and extends to 06^h00^m, $-61°$ and the second starts at 05^h10^m, $-65°$ and extends to 05^h00^m, $-61°$. In the south, an extended ridge is clearly visible in the continuum (Mathewson and Healey, 1964a) from 05^h30^m, $-72°$ to 05^h00^m, $-72°$. A second spur extends to the south-west from 05^h40^m, $-72°$ to 04^h40^m, $-75°$. The emissivity of these arms is about one-quarter that of our local spiral arm.

5. Radio Structure of the Small Magellanic Cloud

The radio structure of the SMC may be divided into two components, the bar and wing component and the discrete source component.

A. THE BAR AND WING RADIO EMISSION

The 1410 Mc/sec and 408 Mc/sec isophotes are co-extensive with the bar and show an extension along the wing of the SMC. The spectral index of the radio emission is about -0.6 between 1410 Mc/sec and 85 Mc/sec but becomes flatter (~ -0.3) between 2650 Mc/sec and 1410 Mc/sec. This may indicate an extended thermal source along the bar which may be due to a number of unresolved thermal sources.

B. DISCRETE SOURCES

At 1410 Mc/sec (Mathewson and Healey, 1964a) and 2650 Mc/sec (Broten, 1969) a number of discrete sources are visible which are identified with emission nebulae in the catalog of Henize (1956). Table IV lists some radio sources in the region of the SMC which should prove interesting for optical and radio studies.

It is most striking that the four non-thermal sources in the wing and bar of the SMC lie on the circumference of the expanding H I shells discovered by Hindman (1967). Two of the non-thermal sources are possibly associated with the emission nebulae N50 and N90 but the other two have no optical identification. These may be extra-Magellanic although one appears to be about 5′ of arc in extent. It is interesting to note that the supernova remnants in the LMC, N49 and N63A, were found on the circumference of a shell of H I which Westerlund and Mathewson (1966) suggested is the remnant of a super-supernova.

In the region to the north-west of the SMC there lies a string of 5 sources with thermal spectral indices. Three of them are extended and at such high galactic latitudes, it seems reasonable to assume that they belong to the SMC. It is suggested that this

TABLE IV

Interesting radio sources in the region of the Small Magellanic Cloud

Source number	Position (1957) R.A. h m	Dec ° ′	Flux Density, 20 cm $\times 10^{-26}$ W m^{-2} (c/sec)$^{-1}$	Size (min. of arc)	Spectra	Remarks
1	00 51.7	72 45	0.6		−0.4	H I Ring, 1 N50, S.N. remnant?
2	01 10.2	73 22	1.0	5	non-th.	H I Ring 1
3	00 21.2	74 37	0.8	4	th.	H I Ring 2
4	00 22.4	74 02	0.6		non-th.	H I Ring 2
5	00 32.5	73 36	0.5		th.	H I Ring 2
6	01 28.6	73 41	0.6		non-th.	H I Ring 3, N90, S.N. remnant?
7	00 14.2	69 55	0.6	6	th.	N.W. extension
8	00 23.2	68 32	1.1		th.	N.W. extension
9	00 38.1	68 41	1.2	10 × 20	th.	N.W. extension
10	00 39.6	71 52	1.0	10	th.	N.W. extension
11	00 54.7	71 14	0.6		th.	N.W. extension
12	00 19.2	76 52	1.1		th.	Southern arm
13	00 41.5	75 55	0.9		th.	Southern arm
14	01 02.9	75 55	0.7	7 × 10	th.	Southern arm, N73
15	01 46.0	74 47	2.5	24 × 45	th.	Southern arm, H I concentration

NW extension is an outer arm of the SMC. Similarly in the south, a line of thermal sources at Dec. $-76°$ may delineate a southern arm.

6. Magnetic Field Structure in the Magellanic Clouds

For reasons discussed previously polarization measurements of the radio emission from the Clouds are very difficult. However, magnetic fields must be present over the optical extent of the Clouds because non-thermal emission is present. From galactic studies (Mathewson, 1968) it appears that the structure of the magnetic field derived from optical and radio polarization measurements agree very closely. Therefore it is reasonable to assume that polarization measurements of stars in the Magellanic Clouds should give a good idea of the general magnetic field structure.

Recently Mathewson, Serkowski and Ford have made extensive optical polarization measurements in the LMC and SMC using the 40-inch and 24-inch reflectors at Siding Spring Observatory. So far only preliminary results are available but it does appear that there is a definite association between the optical and magnetic field structure in the LMC and SMC.

In the LMC, 30 Doradus is the region of highest polarization. It also appears to be the region on which the magnetic field lines focus which is additional evidence that it is the nucleus of the LMC. The magnetic field lines follow the '243 km/sec' arm (McGee and Milton, 1966) out of the 30 Doradus region to the south. They also sweep down from 30 Doradus to the bar which they follow to the end where they turn to the south-west. Visvanathan (1966) had found earlier that the magnetic field was directed along the bar. The highest polarization recorded so far in the LMC is 3.5%.

In the SMC, the magnetic field is directed along the bar and wing. In addition to these magnetic fields associated with optical features, there seems to be in the LMC and SMC, a uniform large scale magnetic field present. The optical polarization vectors delineating this magnetic field are closely parallel to the line joining the two galaxies. This suggests that this magnetic field belongs to the 'Magellanic System', a name given by Hindman *et al.* (1963) to the common envelope of H I in which the LMC and SMC are enclosed.

7. Conclusion

To conclude, I would like to list, firstly, the various investigations in progress which should be published by the end of the year and, secondly, the observations which I would like to see made in the near future.

A. INVESTIGATIONS IN PROGRESS

(1) The 5009 Mc/sec survey of the LMC and SMC using the 210-ft reflector at Parkes by McGee (1969); effective noise level, 0.1 flux-units; half-power beamwidth, 4′ of arc.

(2) The 408 Mc/sec survey of the brighter emission nebulae in the Clouds by Mills and Aller (1969) using the Molonglo Radio Telescope; effective noise level, 0.03 flux-units; half-power beamwidth at the declination of the Clouds, $2\farcm8 \times 3\farcm5$.

The comparison of these two high resolution and high sensitivity surveys at high and medium frequencies should increase tremendously our knowledge of H II regions and supernova remnants – the number of known supernova remnants could easily rise to between 10 and 20.

(3) Epps and Aller (1969) used the Michigan Schmidt telescope at Cerro Tololo to measure the distribution of Hβ in the emission nebulae in the Cloud. They intend to convolve the Hβ isophotes with an appropriate function to make the resolution equal to that of the 408 Mc/sec survey by Mills and Aller. This comparison will be used to study H II regions, supernova remnants and absorption in the Clouds.

(4) The results of the optical polarization measurements of Mathewson, Serkowski and Ford of over 1000 stars in the LMC and SMC should provide a detailed map of the structure of the magnetic field which then can be compared with the distribution of the continuum emission.

B. FUTURE OBSERVATIONS

(1) The measurements of the high frequency spectrum (3000 Mc/sec to 10000 Mc/sec) of the central core of N157A (30 Doradus). It may be non-thermal similar to Sagittarius A. The fact that no hydrogen recombination lines have been detected from N157A suggests that this is quite probable. (In this regard it would be interesting to see if only the 1720 Mc/sec OH line appears in emission in N157A similar to the galactic centre and some non-thermal galactic sources (Goss and Robinson, 1968).)

(2) The measurement of the hydrogen recombination lines from emission nebulae in the Clouds when a more sensitive receiver becomes available at Parkes (McGee, 1969). This would be an excellent method of searching for supernova remnants by measuring the ratio of the line to the continuum temperature (Mezger, 1967; Burke, 1968).

(3) Optical spectra of N94 and N157B in the LMC and N90 and N50 in the SMC to confirm the suggestion based on the radio observations that they are supernova remnants.

(4) Hβ photographs at the positions of the unidentified thermal radio sources listed in Tables III and IV.

(5) The Parkes radio surveys at 1410 Mc/sec and 408 Mc/sec extended to cover larger regions around the Clouds.

(6) Interferometry of the Clouds at 1410 Mc/sec and 2650 Mc/sec using the 210-ft and 60-ft reflectors at Parkes to search for fine structure in the discrete sources.

(7) Accurate spectral index measurements at a number of selected regions in the Clouds.

(8) Low frequency measurements at high resolution to map any extended regions of ionized hydrogen.

References

Broten, N. W.: 1965, *Symposium on the Magellanic Clouds* (ed. by J. V. Hindman and B. E. Westerlund), Mt. Stromlo Obs., Canberra, p. 72.
Broten, N. W.: 1969, private communication.

Burke, B. F.: 1968, *Symposium on* H II *Regions* (ed. by Y. Terzian), W. A. Benjamin Inc., New York, p. 541.
Dickel, H. R., Aller, L. H., and Faulkner, D. J.: 1964, in F. J. Kerr and A. W. Rodgers (eds.), 'The Galaxy and the Magellanic Clouds', *IAU/URSI Symp.* **20**, 294.
Doherty, L., Henize, K. G., and Aller, L. H.: 1956, *Astrophys. Suppl.* **2**, 345.
Epps, A. and Aller, L. H.: 1968, private communication.
Goss, W. M. and Robinson, B. J.: 1968, *Astrophys. Letters* **2**, 81.
Henize, K. G.: 1956, *Astrophys. J. Suppl.* **2**, 315.
Hindman, J. V.: 1967, *Australian J. Phys.* **20**, 147.
Hindman, J. V., Kerr, F. J., and McGee, R. X.: 1963, *Australian J. Phys.* **16**, 570.
Laan, H. van der: 1962, *Monthly Notices Roy. Astron. Soc.* **124**, 125.
Le Marne, A. E.: 1968, *Monthly Notices Roy. Astron. Soc.* **139**, 461.
McGee, R. X.: 1969, private communication.
McGee, R. X. and Gardner, F. F.: 1968, *Australian J. Phys.* **21**, 149.
McGee, R. X. and Milton, J. A.: 1966, *Australian J. Phys.* **19**, 343.
Mathewson, D. S. and Healey, J. R.: 1964a, in F. J. Kerr and A. W. Rodgers (eds.), 'The Galaxy and the Magellanic Clouds', *IAU/URSI Symp.* **20**, 245.
Mathewson, D. S. and Healey, J. R.: 1964b, in F. J. Kerr and A. W. Rodgers (eds.), 'The Galaxy and the Magellanic Clouds', *IAU/URSI Symp.* **20**, 283.
Mathewson, D. S. and Milne, D. K.: 1965, *Australian J. Phys.* **18**, 635.
Mathewson, D. S.: 1968, *Astrophys. J. (Letters)* **153**, L47.
Mathewson, D. S. and Nicholls, D. C.: 1968, *Astrophys. J. (Letters)* **154**, L11.
Mathewson, D. S., Serkowski, K., and Ford, V.: 1969, private communication.
Mezger, P. G.: 1967, in H. van Woerden (ed.), 'Radio Astronomy and the Galactic System', *IAU Symp.* **31**, 229.
Mills, B. Y.: 1959, *Handbuch der Physik* (ed. by S. Flügge), **53** 239–74 Springer-Verlag, Berlin.
Mills, B. Y. and Aller, L. H.: 1969, private communication.
Milne, D. K.: 1969, private communication.
Poveda, A. and Woltjers, L.: 1968, *Astron. J.* **73**, 65.
Shain, C. A.: 1959, in R. N. Bracewell (ed.), 'Paris Symposium Radio Astronomy', *IAU Symp.* **9**, 328, Stanford University Press, Stanford.
Shklovsky, I. S.: 1960, *Soviet Astron.-AJ* **4**, 243.
Verschuur, G. L.: 1969, *Astrophys. J. (Letters)* **155**, L155.
Visvanathan, N.: 1966, *Monthly Notices Roy. Astron. Soc.* **132**, 423.
Westerlund, B. E. and Mathewson, D. S.: 1966, *Monthly Notices Roy. Astron. Soc.* **131**, 371.
Yates, K. W.: 1968, *Australian J. Phys.* **21**, 167.

SATELLITES OF STELLAR SYSTEMS IN GENERAL

ERIK B. HOLMBERG

Uppsala University, Astronomical Observatory, Uppsala, Sweden

This paper gives a summary of some parts of a more comprehensive investigation on the satellites of nearby spiral galaxies. The detailed results will be published elsewhere (Holmberg, 1969).

The investigation refers to 174 more or less prominent spiral systems selected from the catalogue by Holmberg (1964), partly also from the catalogue by de Vaucouleurs *et al.* (1964). It is based on a large number of Mount Wilson 60-inch and 100-inch plates, and on the prints of *The National Geographic Society – Palomar Observatory Sky Survey*, the latter being evaluated down to the practical limit as regards galaxies. From an examination of over 3000 small galaxies in survey areas around the spiral systems, and in nearby comparison areas, a total of about 300 physical companions have been picked out. The survey extends in absolute pg magnitude down to $M = -10.6$.

The satellite groups seem in all respects to be comparable to the Milky Way group and the M 31 group (the two parts of the Local Group), and to the M 81 group. The results are also consistent with previous statistical data available for groups in general. The combined material indicates that the average spiral system has about 8 satellites down to an absolute pg magnitude of -10.6, that the mean separation of satellite from central system is about 110 kpc and the maximum separation 300–400 kpc, and that 30% of the satellites have separations less than 50 kpc.

The survey work on the sky atlas refers to circular areas with a radius of 50 kpc centered on the 174 spiral systems, and to comparison areas of the same size located on either side of each survey area. The latter areas are needed to estimate the number of optical companions. In a survey going down to the limit of the sky atlas there will be considerable disturbances from the background-foreground fields; this is the main reason for keeping the survey areas down to the size mentioned. The work has aimed at picking up all galaxies on the blue prints that have major diameters $\geqslant 1.0$ kpc, as referred to the distance (from photometric data and from redshifts) adopted for the central system; for the most nearby spirals the survey has been extended to a diameter of 0.6 kpc. Magnitudes cannot be directly determined for these small galaxies. The collection of data must be limited to diameters, morphological types (only medium-sized and large objects), separations from the central system, and position angles as measured from the direction of the major axis of this system.

A study of the combined material of physical satellites leads to a number of interesting results. In the first place, the satellites are not distributed at random around those spiral systems that have an edgewise orientation. There is a preferential alignment along the minor axes, position angles of 0°–30° being avoided, which would indicate that a spiral galaxy has no satellites in local latitudes less than 30°. Secondly, there seems to be a correlation between number of satellites and color excess of the

Muller (ed.), The Magellanic Clouds, 109–113. *All Rights Reserved.*

spiral nucleus: the number increases with decreasing color excess. The nuclear color indices have been measured on the Mount Wilson plates, the mean results being $+0\overset{m}{.}38$ (Sc+), $+0\overset{m}{.}62$ (Sc−), $+0\overset{m}{.}74$ (Sb+), $+0\overset{m}{.}76$ (Sb−, Sa), and $+0\overset{m}{.}77$ (So) in the pg–pv system. The excess is defined as observed color *minus* mean color corresponding to the type. Since Seyfert galaxies, as well as other galaxies with abnormal nuclei, have negative color excesses, the correlation indicates a dependence of the number of satellites on the state of the nucleus of the central galaxy. For the third, a correlation is indicated with the hydrogen mass (but not the total mass) of the spiral system; the number of satellites increases with the H I mass. The masses have been estimated from the absolute luminosities and the integrated colors.

It is tempting to try to interpret the above results as indicating that the satellites have been formed from matter, presumably gas, ejected from the nuclear parts of the spiral systems. In low local latitudes the ejection may have been stopped by the gas located in the main body of the spiral, as seems to be the case in M 82. On the other hand, it is difficult to understand and explain the physical ejection mechanism, the condensation processes in the expelled gas, and the large variation that is found in the present material as regards the morphological types of the satellites.

The most important result to be derived from the satellite groups refers to the statistical distribution of absolute diameters (and absolute magnitudes). A comparison between the total number of objects in the groups and the number of *all* galaxies to be expected in the corresponding volume of space indicates that the great majority of galaxies (outside the big clusters) are members of physical groups of the type studied here. Accordingly, the satellite population ought to be highly representative of a given volume of space.

The distribution, on a logarithmic scale, of the absolute diameters (pc) of all the satellites is reproduced in Figure 1. The diameters have been reduced to the diameter system of Holmberg (1958, 1964); the log. diameter, $\log A$, now extends from 2.95 to 4.75. The full curve (large circles) corresponds to the entire material, whereas the dashed curve (small circles) refers to the type group E-So-Ir. Satellites of types Sa-Sb-Sc are represented by the area between the two curves (types can be determined only for objects with $\log A \geqslant 3.55$).

It seems possible to transform the diameter distribution into a magnitude distribution by means of the remarkably high correlation that exists between absolute log. diameter and absolute pg magnitude of galaxies. A study of the data listed in Holmberg (1964) for about 250 galaxies of all types shows that the relation is very pronounced, and apparently linear, the coefficient of correlation being -0.962 ± 0.005 (m.e.). The regression line is represented by $M = -6.00 \log A + 7.14$; the dispersion in absolute magnitude around this line amounts to only 0.40 mag. It should be noted here that the absolute magnitudes are based on a Hubble parameter of 80 km/sec per Mpc. In order to determine the luminosity function for the satellite population we only have to replace the diameter scale at the bottom of Figure 1 with the corresponding magnitude scale. As a check on the accuracy of the transformation, apparent magnitudes have been collected from various sources for all satellites with absolute luminosities higher than $M = -16.5$. The resulting absolute magnitude distribution, as represented by the filled circles in the figure, agrees well with the curve derived from the diameters.

The logarithmic distribution of absolute magnitude for galaxies of types E-So-Ir can be described by a straight line that presumably extends from the lower limit $M = -10.6$ to about $M = -20$; in the interval -20 to -22 the curve seems to fall off rapidly. It is a very interesting fact that this distribution agrees with the curve suggested by Zwicky (1957) for members of galaxy clusters; since the dense clusters are practically free from spirals Sa-Sb-Sc the results ought to be comparable. The agreement also refers to the inclination of the distribution line. A least squares solution, based on the frequencies of Figure 1, gives an inclination of 0.195 ± 0.007, a result that within the limits of the accidental errors agrees with Zwicky's coefficient of 0.2.

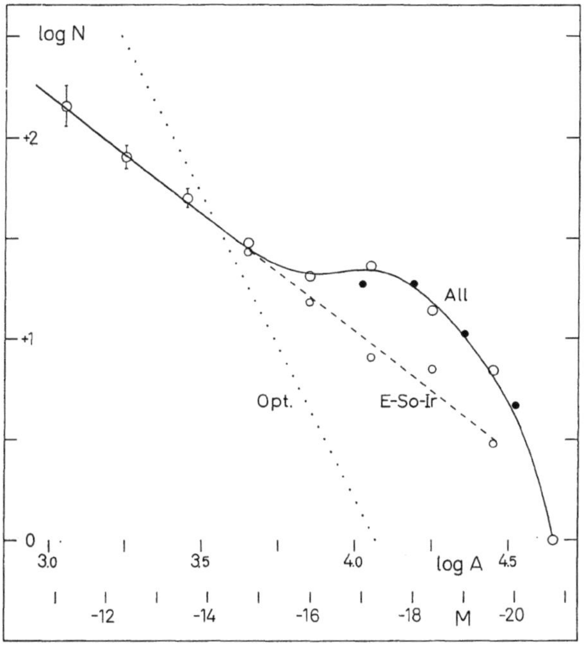

Fig. 1. Statistical distribution of log. absolute diameter (and absolute pg magnitude) as obtained for 274 satellites in 160 survey areas. The large open circles refer to the entire material, and the small circles to types E-So-Ir; the filled circles represent those satellites for which the absolute magnitudes are based on available apparent magnitudes. The dotted line gives the distribution of the subtracted optical companions.

The results obtained indicate that there is, as has been suspected before, a definite lower luminosity limit, $M = -15$, for spiral galaxies Sa-Sb-Sc; the upper limit seems to be near $M = -22.0$. With a good degree of approximation the distribution referring to this group can be described by a normal error-curve, with a mean absolute pg magnitude of -17.7 and a dispersion of 1.2. mag. It may be recollected that, except for a systematic displacement in magnitude, this distribution agrees rather well with the classical luminosity function derived by Hubble (1936) from a study of the brightest resolved stars; Hubble's material referred mainly to Sc and Sb spirals.

In order to make possible a more detailed study, the distributions referring to the brighter magnitude classes have been reproduced in Figure 2. The full curve (large circles) and the dashed curve (small circles) correspond to the curves of Figure 1. An absolute calibration has been introduced, giving the number of galaxies per magnitude class in 1 Mpc^3. The calibration is based on a space density of 0.17 galaxies brighter than $M = -15.0$ per Mpc^3, a mean result that has been derived by the writer from analyses of the magnitudes, diameters, and redshifts of galaxies.

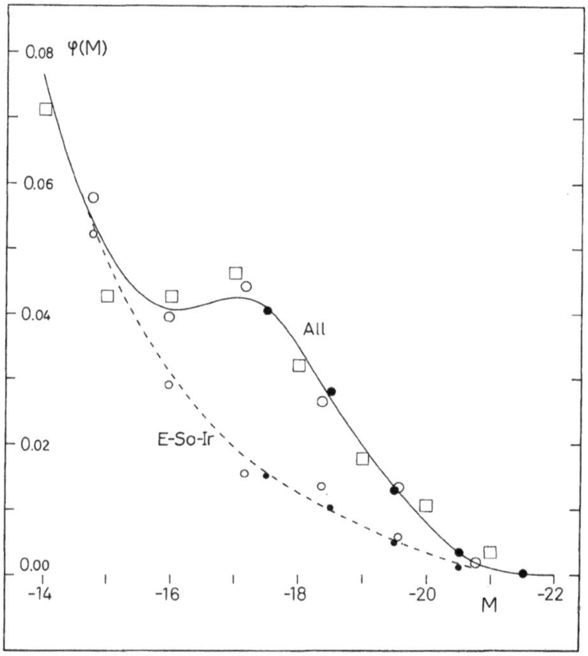

Fig. 2. The brighter end of the luminosity function (corresponding to the right half of Figure 1), as calibrated for 1 Mpc^3. The filled circles give the distributions derived for all galaxies with known redshifts north of galactic latitude $+30°$. The squares refer to the Local Group and the M 81 group.

In the second figure two interesting comparisons are made with results available from other sources. The squares represent the Local Group and the M 81 group, a material that is probably complete down to $M = -13.5$. The class frequencies are, except in the faintest class (5 galaxies), overlapping means: 50% from the central class and 25% from each adjoining class. In spite of the small total number, only 18, there is a good agreement with the full curve. The large full circles represent the distribution derived for all the galaxies north of galactic latitude $+30°$ (Virgo cluster omitted) in the available redshift lists, whereas the small full circles refer to the E-So-Ir galaxies alone. It is very satisfactory to find that the results from the redshift material are in perfect agreement with the results obtained from the satellite groups, especially in view of the fact that the former results are based on a much larger population as regards galaxies of high luminosity.

To summarize the above results, it can be stated that the pg luminosity function ($H = 80$) corresponding to all galaxies in 1 Mpc^3 (outside the big clusters) is described by the formula

$$\varphi(M) = 40 \times 10^{0.195M} + 0.025 \cdot e^{-0.35\,(M+17.7)^2},$$

where the first term refers to types E-So-Ir and the second term to types Sa-Sb-Sc. The formula represents the distribution from $M = -10.6$ to $M = -19$ a -20; beyond this limit the curve rapidly approaches zero. Whereas the total number of galaxies is 0.17 per Mpc^3 for magnitudes brighter than -15.0, it increases to about 0.8 for $M \leqslant -10.6$.

References

Holmberg, E.: 1958, *Medd. Lunds Astron. Obs., Ser. II*, No. 136.
Holmberg, E.: 1964, *Medd. Uppsala Astron. Obs.*, No. 148 = *Arkiv Astron.* **3**, No. 30.
Holmberg, E.: 1969, *Medd. Uppsala Astron. Obs.*, No. 166 = *Arkiv Astron.* **5**, No. 20.
Hubble, E.: 1936, *Astrophys. J.* **84**, 279 = *Contrib. Mount Wilson Obs.*, No. 548.
Vaucouleurs, G. and A. de: 1964, *Reference Catalogue of Bright Galaxies*, The University of Texas Press, Austin.
Zwicky, F.: 1957, *Morphological Astronomy*, Springer-Verlag, Berlin.

PART III

CURRENT THEORETICAL APPROACHES

PHYSICAL PARAMETERS OF SUPERGIANTS IN THE MAGELLANIC CLOUDS

TH. and J. WALRAVEN
University Observatory, Leiden, The Netherlands

Abstract. Five-colour photometry *V, B, L, U* and *W* permits to distinguish members of the Magellanic Clouds from the galactic foreground stars. The colour-indices of members are identical with those of galactic supergiants, if corrected for reddening.

The members of the Magellanic Clouds show a wide variety in brightness and spectral type, and are not much affected by reddening.

The colours of this two-dimensional array are compared with the model atmosphere calculations of de Jager and Neven (1967) and Mihalas (1966).

1. Introduction

While intrinsic colours of main-sequence stars are well known, the colours of the supergiants are vague and uncertain, due to their relative scarcity, unknown distance and strong interstellar reddening.

Only in the Magellanic Clouds where we observe such stars in abundant number and variety, all at about equal distance, which is known with reasonable certainty, and little affected by interstellar reddening, intrinsic colours can be studied.

We observed about 600 stars in the Magellanic Clouds, with the five-colour photometer at the Leiden Southern Station.

By the use of the five spectral bands it appeared possible to separate the galactic foreground stars from the supergiants which are members of the Clouds.

It turned out, that about 200 stars were not members of the Clouds. These are in majority type F and G stars, but some B- and A-type stars were observed among them.

The most striking feature of the members is a steady increase in colours towards the red, with increasing brightness. At medium and later type B, the brightest supergiants are conspicuously redder than the main-sequence stars. By some investigators this has been ascribed to interstellar reddening.

In the present paper we demonstrate, that this effect is intrinsic, since it is predicted by model calculations.

2. The Observations

When we discovered in 1960, that the five-colour photometer enabled us to distinguish galactic foreground stars from members of the Clouds, and arrived at the conclusion that practically all stars of type F and G, which at that time were still considered as possible members, are actually foreground stars, we undertook a survey of as many as possible stars, in the hope to obtain a better insight in the statistical distribution of the colours and brightness of the stars in the Magellanic Clouds.

We made use of the results of a similar investigation conducted spectroscopically by the Radcliffe astronomers Feast *et al.* (1960), who had selected already a considerable

Muller (ed.), The Magellanic Clouds, 117–135.

number of the brightest members. In the Large Cloud use was made of the HDE catalogue, from which O- and B-type stars had been selected and for which finding charts had been prepared by Dr. K. K. Kwee of the Leiden Observatory. Also a few F-type stars discovered by Fehrenbach were observed.

For fainter stars no data were available and we simply inspected one star after another, using short integration times of one minute only. In this way we probed everywhere the field of the Small Magellanic Cloud.

The colour effects are so large, that already during the observations we could decide about membership with some certainty. Thereby we could avoid going too far in the outer regions, where few members occurred. Probing in this way the outer regions we encountered an association surrounding HD 7583, R.A. $01^h12^m24^s$, Dec. $-73°32'.8$ (1960), lying in an extension of the Small Cloud towards the Large Cloud. Beyond this group the relative numbers of supergiants decreased, although we had the impression, that more intense searching might reveal supergiants further away from the Small Cloud.

We also noted the fact, that in other outlying regions of the Small Cloud, only fainter stars were members, but usually of more advanced spectral type.

The Large Cloud seemed different in the sense, that the faint stars are more systematically of early type, and are very abundant, so that we confined systematic probing to a small region lying near the edge of the Cloud and surrounding the stars HDE 271191, R.A. $05^h21^m32^s$, Dec. $-65°47'.1$ (1960) and HDE 271192, R.A. $05^h21^m33^s$, Dec. $-65°54'.3$ (1960). In this area practically no stars of late type B were encountered. Thus, at least locally, considerable differences do occur in the distribution of spectral type.

The faintest stars observed are of the 15th magnitude a limit which is set by the size of the diaphragm, which is 15 sec of arc in diameter.

For some of the 15th magnitude B stars we made repeated observations with longer integration times in the hope to be able to apply our luminosity criteria, such as found from the galactic main-sequence stars (1959).

These more accurate observations revealed, that we had indeed reached the main-sequence stars, but of a type not later than B 1, for which the luminosity effects are still small. But the colours, although in agreement for the same star, varied considerably for different stars. It was found later, after examining large scale photographs made by A. R. Hogg, with the Mount Stromlo 74 inch, that many of the stars were disturbed by neighbours.

We must consider our observations as referring mainly to supergiants.

3. The Colour-Colour Diagrams

A list of all observed stars will shortly be published in *Astronomy and Astrophysics*. Colour-colour diagrams are shown in Figure 1. The left-hand upper part shows stars of the association in the Large Cloud surrounding HDE 271191 and HDE 271192. These are faint stars of early type, partially reddened. Quite different is the distribution in the left-hand lower diagram, which shows stars of the Small Magellanic Cloud. It is seen, that the number of stars which have larger Balmer discontinuity, which have spectral type A or F, has increased relative to the number of early B stars.

The stars fall inside a broad band, which after reaching a large Balmer-jump, bends towards redder colours, where cepheids are observed. The colours of some cepheids are indicated with crosses. However the sudden decrease of non-variable stars, after reaching the bend, is notable.

In the lower righthand part of the figure the same band of supergiants is outlined by dashed lines and the large separation of F- and G-type foreground stars from this band may be noted. In the righthand upper part the brighter supergiants of the Large Magellanic Cloud (mostly HDE stars) is shown. The main-sequence line is indicated in all the diagrams.

The band for supergiants runs parallel to the main-sequence for B-type stars and crosses it at about type A 3. A confusion is excluded in this case, when use is made of the colour-colour diagrams $B-L$ vs $V-B$ and $U-W$ vs $V-B$.
Confusion between members and foreground stars of late spectral types is possible in the region to the red side of the dashed-dotted line indicated in the right-hand lower part of Figure 1.

4. Magnitude-Colour Diagram

Plotted in a magnitude-colour diagram (Figure 2) the stars occupy a broad band, starting at the fifteenth magnitude, rising upwards and bending towards the red, resulting in a long tail of tenth magnitude stars in the case of the Large Magellanic Cloud.

This tail of very luminous supergiants of type F and G is absent in the Small Magellanic Cloud. The brightest supergiants of type G in the Small Cloud are cepheids.

For comparison in the diagram the mean path of the stars of the double cluster h and χ Persei is shown, as a dashed line. Although the typical bending to the red is shown by this path, it differs from that of the Magellanic Cloud supergiants.

In the Small Cloud we observe much more stars of the thirteenth magnitude lying under the horizontal part of the path of h and χ Persei, in the Large Cloud, many stars between the tenth and eleventh magnitude are lying above the h and χ Persei line.

In the diagrams we inserted the lines of constant bolometric magnitude, shifted vertically such as to pass through the brightest stars of the Clouds. Especially in the Large Cloud this line reveals the existence of an upper limit to the luminosity of the stars, such as was first suggested by Feast *et al.* (1960). According to the diagram only one star in the Large Cloud surpasses this limiting magnitude, this is HD 32228, a Wolf-Rayet star which according to Underhill (1968) reveals in its spectrum that in reality a close group of stars is observed.

The remarkable star HD 33579, $m_v=9.14$, which is probably the visually brightest known supergiant, falls on the line.

The line of constant bolometric magnitude may touch also the known stars of type

Of, one in the Large Cloud HDE 269698, Rad. no. 115, $m_v = 12.25$, Sp. O5f and in the Small Cloud, Rad. no. 13, $m_v = 12.26$, Sp. O: f:.

In Figure 2 these stars can be found as the bluest stars at $m_v = 12.2$.

The remarkable separation in magnitude of the Of stars from fainter stars of the same blue colour, can be noted in both Clouds. These fainter blue stars lie along the main-sequence line as indicated on the diagram.

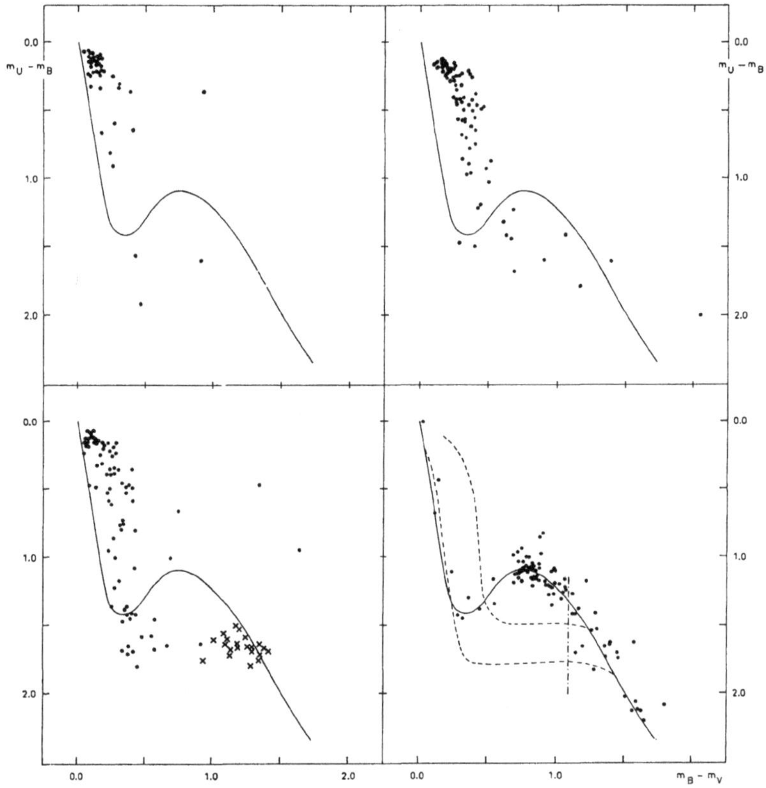

Fig. 1. Colour-colour diagrams. Upper left: association in LMC. Upper right: general field of LMC. Lower left: general field SMC (crosses denote cepheids). Lower right: galactic foreground stars. Scales in magnitude with O stars at origin.

During the observations red stars were avoided by comparing the visual telescopic image, with a blue photograph, but we estimate from the distribution of the foreground stars (see Figure 1), that only to the right side of the vertical dashed line, a selection effect has been active.

The lack of yellow faint stars to the left side of the dashed line, we consider as a real characteristic of the magnitude-colour array.

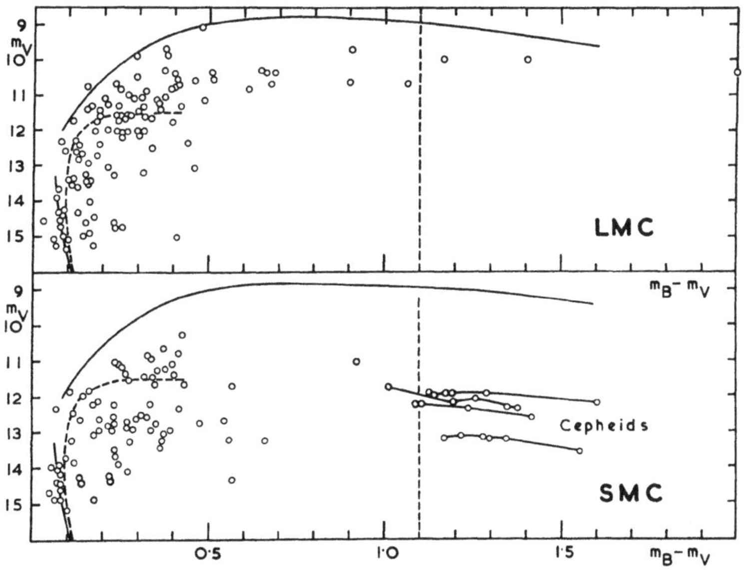

Fig. 2. Magnitude-colour diagrams. Abscis in magnitudes, with zero-point at unreddened O stars. Ordinate in visual magnitude. Dashed line: mean path h and χ Persei. Upper end of main-sequence and limiting constant bolometric magnitude shown as drawn lines.

5. The Intrinsic Colours

Both in the colour-colour diagrams and in the brightness-colour diagrams the stars occupy strips, which are roughly $0^m.4$ wide in the direction of the colour-index $m_B - m_V$. At first sight this might be due to interstellar reddening, but a careful analysis has shown, that this is not the case, but that the variations in colour $m_B - m_V$ are intrinsic and correlated with luminosity and spectral type.

This was discovered by us a short time after the observations were made and preliminary results were given at the IAU meeting at Berkeley (1961), the *IAU-URSI Symp.* **20** at Sydney (1963) and for the later type supergiants at the *IAU Symp.* **24** at Saltsjøbaden (1964). According to these discussions the colours of the Magellanic Cloud supergiants vary systematically with luminosity, becoming redder with increasing luminosity.

A rather surprising effect was found in the Balmer-jump.

According to theory the maximum of the Balmer-jump should increase for stars of increasing luminosity. This effect is confirmed by the observations, while going from main-sequence stars to the supergiants. But contrary to expectations the Magellanic Clouds supergiants themselves, i.e. going from class Ib to class Ia, showed a decrease of the maximum of the Balmer-jump and this was confirmed by observations of galactic supergiants. In other words the Balmer-jump passes through a two-dimensional maximum at class Ib, type F 0; a typical example we find in α Lep, F 0 Ib. Spectral classifications of Magellanic Cloud stars of this type do not yet exist.

For class Ia stars the maximum value of the Balmer-jump is reached at spectral type F 8 and is considerable smaller. A galactic supergiant representing this case is δ C Ma, F 8 Ia. If we restrict ourselves to spectral type A 0, the Balmer-jump increases from class V to class IV and steadily decreases while going from class II to more luminous stars.

We suggested tentatively at the Sydney Symposium, that this effect could be explained, if it is assumed, that with increasing luminosity, the contribution of electron scattering to the opacity of the atmosphere becomes increasingly more important, i.e. not only for the hottest stars, but also for the cooler stars, with extremely rarified atmospheres.

In recent years a considerable number of model atmosphere calculations have been published. We noted, that the models computed by de Jager and Neven, which cover a particularly wide range of parameters, agree remarkably well with the observations of the Magellanic Clouds. They strengthen the evidence, that the atmospheres of the most luminous stars are dominated by electron scattering.

De Jager and Neven's calculations consist of two parts: the first set of models, designated as the fifty models (1957) do not clearly show the systematic reddening with decreasing $\log g$, but this effect is well pronounced in the second set of models, designated as the sixty models (1967). They differ from the fifty models with respect to the variation of temperature with optical depth. The sixty models show a systematic increase of colours for diminishing $\log g$, which of course reflects the observed reddening as a function of luminosity. We noticed also, that in the colour-colour diagrams derived from the models, the lines of constant $\log g$ showed curves and humps, which seemed present, also in the observations, but which we had smoothed out in the previous discussion, as they seemed unreal to us. We therefore subjected the observations to a renewed discussion.

The stars were arranged in groups of magnitude V, in steps of 0.5 magnitude. For each group colour-colour diagrams were plotted. Through the plots freehand lines were drawn, not as previously through the points, but following closely the bluest stars. Since the number of stars in each group is not large, we did of course not draw a line which filled everywhere the gaps and empty spaces, but avoided humps and curvatures, which seemed unreasonable as compared to the model curves. The resulting blue envelope lines we then assembled in one composite diagram, where they produced a rather systematic set of curves, which showed the well known shift in colour with increasing magnitude. At only a few places the lines of successive groups touched or crossed each other. In these cases we corrected the shapes in order to obtain a more systematic system of curves, but only making changes towards bluer colours. Such corrections amounted to not more than a few hundreds of a magnitude, which is small compared to the total reddening, which is roughly $0\overset{m}{.}25$, from the 10th to the 14th magnitude. The corrected curves, which we now designate as blue envelope lines are shown in Figure 4, for $[B-U]$ versus $V-B$ and in Figure 5 for $[B-L]$ versus $V-B$, and for $[U-W]$ versus $[B-U]$ in Figure 6. In these and further diagrams we use systematically logarithm of intensities instead of magnitudes.

The quantities between square brackets are reduced colours, which can be derived from the observed colours from the formula

$$\begin{aligned} [B-U] &= B-U-0.63\,(V-B)+0.0550 \quad &(\mathrm{a})\\ [B-L] &= B-L-0.42\,(V-B)+0.0240 \quad &(\mathrm{b})\\ [U-W] &= U-W-0.44\,(V-B)+0.0050 \quad &(\mathrm{c}) \end{aligned} \tag{1}$$

This transformation consists of a deformation of the diagrams, such, that the reddening lines for O stars become horizontal. The zero-points of the scales $[B-U]$, $[B-L]$ and $[U-W]$ now refer to O-type stars as origin. The horizontal $V-B$ scale is unchanged and shows the blue envelope lines as they are observed in the Magellanic Clouds, i.e. not corrected for reddening by galactic foreground dust. We estimate the effect of galactic foreground reddening as 0.020 in the logarithmic scale or $0^{\mathrm{m}}.05$ if expressed in magnitudes. This quantity is indicated by blue stars at high galactic latitude as well as by main-sequence stars of the Magellanic Clouds.

The model computations have been processed as follows. From the logarithm of the flux as given in the tables of de Jager and Neven (1967) we subtracted the logarithm of the flux of the model with $T_e = 41\,600°$ and $\log g = 1$. This was done to simulate the procedure followed for the observations where O stars were chosen as zero-point. The relative fluxes were interpolated at the effective wavenumbers of the five spectral bands of our photometry V, B, L, U and W. The effective wavenumbers are given in Table I.

TABLE I

Effective wavenumber (μ^{-1}) for unreddened O star in zenith

V	B	L	U	W
1.850	2.336	2.615	2.765	3.085

Next the relative fluxes at B, L and U were corrected for the influence of the hydrogen lines, since these were not included in the model computations. For this correction a simple formula was introduced, which consists of assuming, that every hydrogen line from Hδ onward has the same equivalent width (expressed in wavenumbers). The lines, which increase in number towards the Balmer limit take away from the continuous spectrum a fraction, which rapidly increases towards the Balmer limit. At a certain point in the spectrum the mean residual intensity becomes equal to that of the Balmer continuum further on. In our model we represent the intensity distribution by a system of hydrogen lines of equal width smoothed out to one continuous curve and we let the Balmer continuum start at the place, where the smoothed curve of the hydrogen lines reaches the same residual intensity as present in the real continuum.

We checked the accuracy of this approximation by comparing the results in cases, which were also computed exactly, based on the precise model computations of the hydrogen lines of Mihalas.

For the use of this simple but effective approximation we need the knowledge of the strength of the hydrogen line Hδ and the Balmer-jump D. In the case of the de

Jager and Neven models the Balmer-jump D is given directly as the difference in the logarithm of the flux, just before and just below the Balmer limit. For the strength of the hydrogen line, Hδ, computations have been made by de Jager and Neven, but it appeared, that for the strong lines the results are not conform to the observed values. In this respect the strength of Hδ as computed by Mihalas agrees better. But unfortunately, the computations of Mihalas do not cover the wide range of parameters we need for the study of the Magellanic Cloud supergiants.

As a compromise we have chosen the strength of Hδ as given by Mihalas, complemented with the results of de Jager and Neven for the weaker lines. The adopted strength of the hydrogen lines is illustrated in Figure 3.

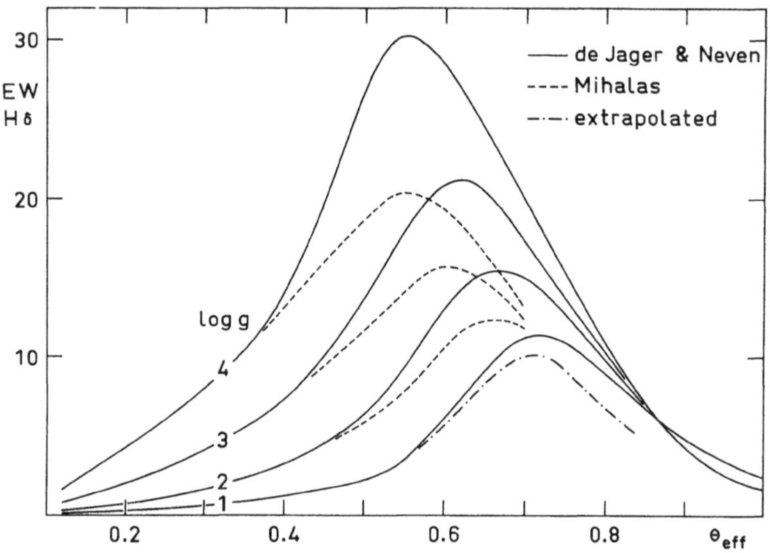

Fig. 3. Model computations of equivalent width of Hδ.

The effect of the correction for hydrogen lines on the fluxes from the models is small for band B and U, therefore the colour-colour diagram of $[B-U]$ versus $V-B$ is not influenced very strongly by the correction. On the other hand the character of the colour-colour diagram $[B-L]$ versus $V-B$ is completely determined by the strength of the hydrogen lines. And to such an extent, that one may consider the model curves for constant $\log g$ in Figure 5 simply as a distorted version of the lines for constant $\log g$ in Figure 3.

We have combined, in Figure 4, the observed blue envelope lines and the model curves for $\log g$ is constant and shifted the latter such, that the line for $\log g = 4$ coincides with the intrinsic main-sequence line, which itself was shifted into a position corresponding to that of the Magellanic Cloud stars, i.e. reddened by $0\overset{m}{.}05$.

In the direction of the scale $[B-U]$ no corrections were applied.

In Figure 4 are moreover shown black-body colours represented by the straight line at the bottom of the picture. The points of constant temperature of the models were connected by dashed lines, which we prolongated tentatively towards the black-body colours of the same temperature. We constructed a similar combination for the diagrams $[B-L]$ vs $V-B$, shown in Figure 5, where in the $V-B$ scale the same correction was applied to the models as in Figure 4, while the direction of the scale

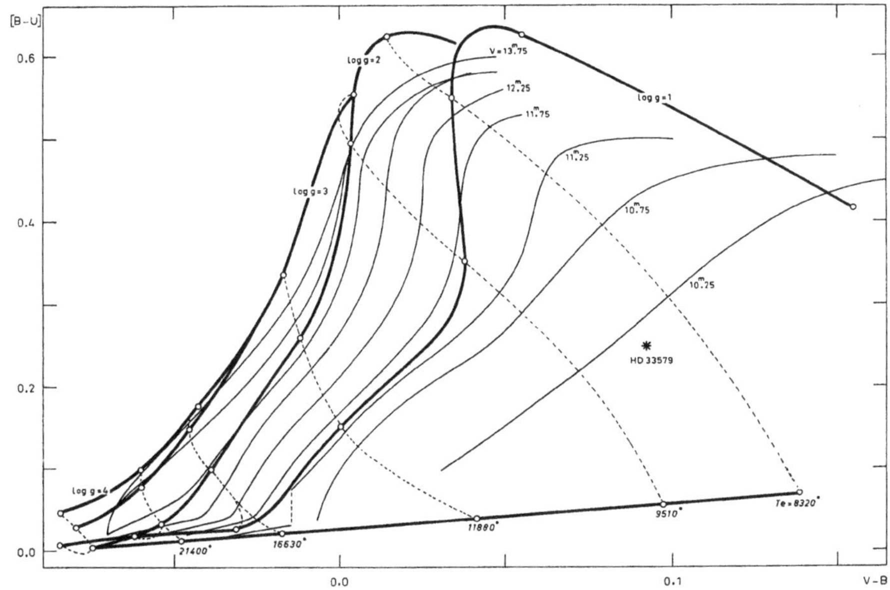

Fig. 4. Superposition of models and observations. Thick lines: (models) log g is constant. Thin lines: (observations) blue envelope lines for V is constant. Dashed lines: (models) effective temperature is constant. Black body colours: indicated by straight thick line near bottom. The $V-B$ scale is shown as observed in the Clouds, i.e. not corrected for galactic foreground reddening.

$[B-L]$ a shift of $0\overset{m}{.}025$ was applied to the models. This shift seems reaonable, as by using it we take account of the fact, that the O stars, which have served for the determination of the zero-point of the observed colours, have hydrogen lines with a strength of somewhat less than 1 Å.

In the figure the black-body colours and lines of constant effective temperatur are also inserted. In the diagrams the position of the most luminous supergiant in the Large Cloud HD 33579 is indicated. In Figure 6, we show the model curves in the colour-colour diagram $[U-W]$ versus $[B-U]$. It appeared, that a considerable discrepancy exists between the models and the observations, the reason for this is not yet understood. We suspect, that somehow the contribution of the Rayleigh scattering is underestimated, since this is the only source of opacity which increases rapidly towards shorter wavelength. It appeared, that the models could be brought into reasonable agreement with the observations, both for the main-sequence stars

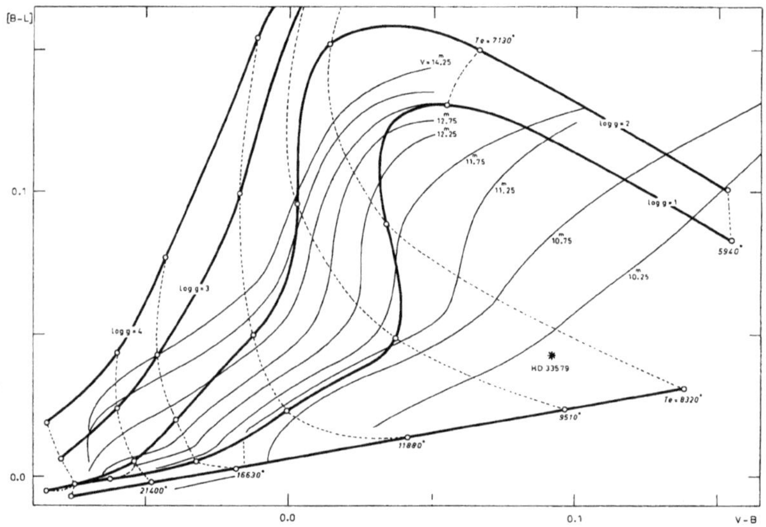

Fig. 5. Superposition of models and observations. Thick lines: (models) $\log g$ is constant. Thin lines: (observations) blue envelope lines for V is constant. Dashed lines: (models) effective temperature is constant. Black body colours shown as thick line near the bottom. The $V-B$ scale is shown as observed in the Clouds, i.e. not corrected for galactic foreground reddening.

and the supergiants by increasing $[U-W]$ of the models with a factor 1.4 and subtracting 0.010. The result is shown in the lower part of Figure 6, where the dashed line represents as typical main-sequence the Orion stars, the shaded area, galactic bright field stars, which are on the average somewhat older than the Orion stars. The logarithm of gravity for these stars varies between three and four. The thin lines incidate again the intrinsic lines for the Magellanic Cloud stars. The exceptional position of HD 33579 may be noted.

As can be seen in the diagrams 4, 5 and 6 the progressive change in colour with increasing luminosity of the stars is shown by the models as well and reflects simply the decrease of effective gravity. A comparison of Figure 4, $[B-U]$ versus $V-B$, and Figure 5, $[B-L]$ versus $V-B$, strengthens our opinion that the variations in colour $V-B$ are intrinsic and not due to reddening by interstellar dust.

From the systematic displacement of the blue envelope lines in itself it follows already that interstellar dust cannot be responsible.

However the effect could still be ascribed to circumstellar dust present in increasing quantities around more luminous stars.

Should we attribute this effect to circumstellar reddening and correct the lines of the bright stars by shifting them towards the blue, into a position corresponding with that of the fainter stars, the same displacement necessarily should be made in both diagrams. In Figure 5 this would lead to the conclusion, that the strength of the hydrogen lines would be about the same for all luminosities. Nobody would accept this.

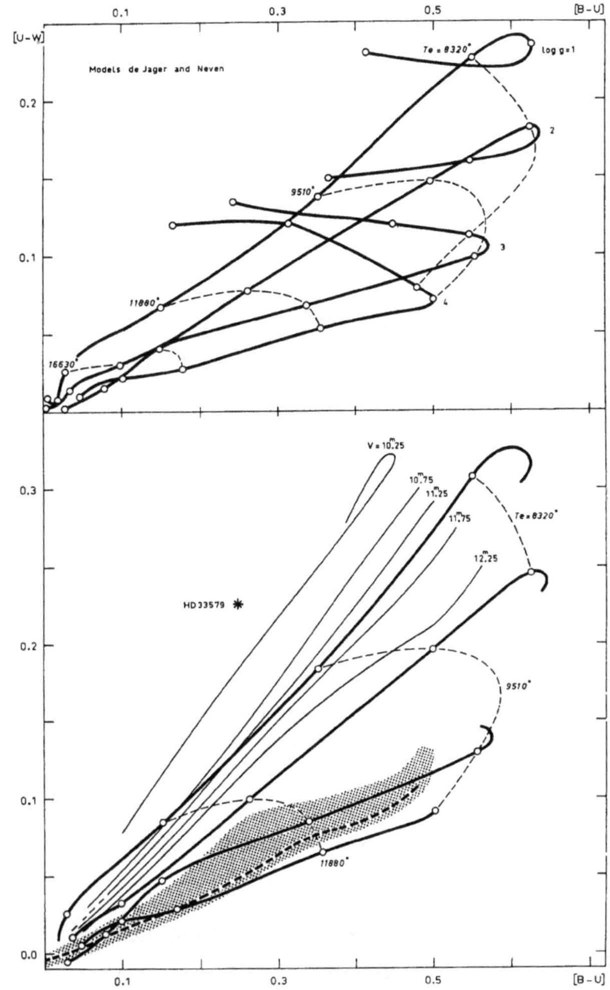

Fig. 6. Colour-colour diagrams $[U-W]$ versus $[B-U]$. Above; (models), thick lines: $\log g$ is constant. Dashed lines: (models) effective temperature is constant. Below; thick lines: (models modified) $\log g$ is constant. Thin dashed lines: (models modified) effective temperature is constant. Thick dashed line: (observations) main-sequence from Orion stars. Shaded area: (observations) main-sequence from galactic field stars. Thin lines: (observations) Magellanic Cloud stars, V is constant.

Obviously we must consider the systematic redshift with increasing brightness as an intrinsic effect.

Moreover it should be mentioned that the diagram in Figure 5 being determined entirely by the strength of the hydrogen lines, is in first instance something different from the diagram in Figure 4, which reflects the behaviour of the continous spectrum. Nevertheless the values of $\log g$, which we can read off in both diagrams for the various lines of constant observed brightness agree closely, as can be seen in Table II. From

TABLE II

$\log g_{eff}$

V_J		10.75	11.25	11.75	12.25	12.75	13.25	13.75	14.25
T_{eff} 21400°	$[B-U]$					1.75	2.34	3.20	3.40
	$[B-L]$					2.00	2.64	3.23	3.53
T_{eff} 16630°	$[B-U]$			1.18	1.46	1.70	2.12	2.80	3.00
	$[B-L]$			1.15	1.46	1.70	2.14	2.72	3.00
T_{eff} 11880°	$[B-U]$	0.75	0.98	1.14	1.37	1.66	1.86	2.24	2.50
	$[B-L]$	0.90	1.10	1.25	1.44	1.66	1.90	2.15	2.32
T_{eff} 9510°	$[B-U]$	0.56	0.72	1.06	1.30	1.56	1.80	1.94	2.00
	$[B-L]$	0.87	0.96	1.15	1.33	1.58	1.76	1.90	2.02
T_{eff} 8320°	$[B-U]$		0.30	0.68	0.90	1.00	1.10	1.15	1.15
	$[B-L]$		0.60	0.92	1.14	1.28	1.36	1.45	1.58

this close agreement we may conclude, that the model computations of de Jager and Neven (1967) produce spectral distribution with a high degree of internal consistency.

We now go a step further and see how well the effective gravity given in Table II agrees with the value one would expect.

The gravity has been computed from radius and mass as follows. For the mass we use the mass-luminosity relation

$$\log \mathcal{M}/\mathcal{M}_{\odot} = -0.14\, M_{\mathrm{bol}} + 0.19 \qquad (2)$$

which was derived from the results of the calculations of evolutionary models given by various investigators. The absolute magnitude we derived using a distance modulus for both Clouds, $m-M=18.75$.

The bolometric correction can be found directly from the flux tabulated by de Jager and Neven. We denote the Newtonian gravity computed from mass and radius as g_{New} in Table III.

TABLE III

Newtonian gravity, effective gravity and radiation pressure

V_J			10.75	11.25	11.75	12.25	12.75	13.25	13.75	14.25
M_V			−8.0	−7.5	−7.0	−6.5	−6.0	−5.5	−5.0	−4.5
$T_{eff}=16630°$	g_{New}	cm/sec²			242	327	441	594	802	1080
	g_{rad}	cm/sec²			55	65	78	112	195	257
	g_{eff}	cm/sec²			14.6	28.8	50.1	151	562	1000
	ξ	km/sec			64	53	46	27	5	
$T_{eff}=11880°$	g_{New}	cm/sec²	55.3	74.6	100.7	135.8	183.2	247	333	452
	g_{rad}	cm/sec²	19.5	23.4	26.9	32.4	41.7	54	74	91
	g_{eff}	cm/cse²	6.6	11.0	15.8	25.1	45.7	76	151	234
	ξ	km/sec	33	30	30	28	22	19	13	12
$T_{eff}=9510°$	g_{New}	cm/sec²	9.5	39.8	553.7	72.4	97.7	131.8	177.8	240
	g_{rad}	cm/sec²	10.2	11.5	15.0	18.8	24.6	31.6	37.2	43
	g_{eff}	cm/sec²	5.2	6.9	12.6	20.9	37.2	60.3	81.3	105
	ξ	km/sec	23	25	20	18	14	11	12	13

We furthermore computed the counter-gravity $g_{\rm rad}$ due to the force exerted by the radiation passing through the atmospheric layers as:

$$g_{\rm rad} = -\frac{\bar{\kappa}\sigma}{c} T_{\rm eff}^4 \tag{3}$$

where $\bar{\kappa}$ is the Rosseland mean opacity.

As can be seen in Table III the sum of $g_{\rm eff}$ and $g_{\rm rad}$ is not yet equal to $g_{\rm New}$, but is systematically proportional to $g_{\rm New}$ in a fairly wide range. A possible explanation of the discrepancy might be, that the effect of turbulence has not been taken into account. We tentatively computed the values of the turbulent velocity ξ which would bring the effective gravity into agreement with the computed values of $g_{\rm New}$ and $g_{\rm rad}$ from the formula:

$$\frac{g_{\rm New} - g_{\rm rad}}{g_{\rm eff}} = 1 + \xi^2 \frac{\bar{\mu}}{2RT} \tag{4}$$

where $\bar{\mu}$ is the mean molecular weight for which we adopted 0.80, and R is the gas constant.

The values for ξ are shown in Table III.

They are of the same order as the macro turbulence but larger than the micro turbulence observed in supergiants. The computed velocities exceed the speed of sound and this seems an obstacle to the conclusion that turbulence explains satisfactorily the low value of $g_{\rm eff}$ since complications arise in the problem of energy transport, temperature equilibrium and so on. The observation that most supergiants show at least Hα in emission proves that the real situation is more complicated than being simply that of an atmosphere in hydrostatic equilibrium.

Thus we must conclude that the value of $g_{\rm eff}$ varies such as one would expect from mass and radius but for some unknown reason is systematically somewhat too low.

A further check of the correctness of the models is the comparison of the observed strength of Hδ and the strength deduced from $[B-L]$. The observations of Hδ for supergiants are not very numerous but sufficient to show that the agreement is good.

Let us summarize the results of the comparison of our observations with the models.

There was a discrepancy in scale of $[U-W]$. This might be caused by incorrectly taking into account the Rayleigh scattering and partly by the effect of line blanketing in the ultra-violet. In both cases corrections will affect mainly the ultra-violet band W and have very little effect on the other parts of the spectrum. Figure 6 shows, that at least qualitatively i.e. after changing the scale of $[U-W]$, there is an agreement with the other diagrams.

Another discrepancy was in the strength of the hydrogen lines, where de Jager and Neven made the strong lines far too strong and Mihalas found more realistic values. However this problem has no bearing on our case, since we are only discussing stars with weak hydrogen lines.

A quite satisfactory agreement was found in all other respects, i.e. the agreement between the colour-colour diagrams $[B-U]$ respectively $[B-L]$ vs $V-B$, the qualitative agreement of g_{eff} with the value derived from mass and radius and the good agreement between $[B-L]$ and the observed strength of the hydrogen lines. We therefore conclude, that the model computations have been rather successful and that we may attach a significant meaning to the values of gravity and effective temperature, which they indicate for the observed stars. On the other hand, they proved the reality of the systematic progress of colour with increased luminosity, which we must consider as intrinsic properties of the supergiants. It seems reasonable to assume, that the blue envelope lines we have derived, represent the intrinsic colours for supergiants.

6. The Physical Parameters

It follows from the comparison, that many of the supergiants fall in the region between $\log g=1$ and $\log g=2$. Towards lower temperatures all supergiants seem to fall beyond $\log g=1$. However we cannot trust the comparison at these low temperatures, since the metal-line blanketing which is not taken into account in the models, becomes important here.

The most remarkable result that follows from the comparison, is the extremely low gravity for the brightest stars, which is obviously unexpected, since there exists no model computations for these extreme low gravities. As can be seen in Figures 4 and 5, the model-lines of $\log g=1$ and $\log g=2$ for the higher temperatures closely approach the black-body line. The observed lines of constant brightness behave precisely in the same way. From an inspection of the tables of the model computations it is immediately clear, that for such models, which lie close to the black-body line, the opacity is constant throughout the depth of the atmosphere and has systematically the same value, i.e. that of electron scattering.

The diagrams suggest very strongly, that a model curve for still lower gravity e.g. $\log g=0$, would follow the black-body line still more towards lower temperatures. At approximately 13000° it may rise, due to the appearance of a Balmer-jump. At still lower temperatures, say about 9000°, one would expect the Balmer-jump to rise more rapidly from analogy with the line $\log g=1$.

The observed line however does not show this, which indicates, that the negative hydrogen absorption and metal-line blanketing become predominant, before the Balmer-jump could increase. We must therefore conclude, that the electron scattering plays an important rôle, with regard to the shape of the intrinsic lines. It has the tendency to push the colours towards the black-body line, by decreasing the effect of the free-bound transitions of hydrogen, ever more so for the lower gravities.

7. Spectral Types

We have made use of the spectral classification of Feast, Thackeray and Wesselink (1960) to locate the mean position of various spectral types in the diagrams. Very similar results were obtained by means of spectral types of galactic supergiants. The latter stars are in general much reddened by interstellar dust and we used their position in the diagram $[U-W]$ vs $[B-U]$ to estimate their intrinsic colours. The results are shown in Figures 7 and 8. The position of the spectral types is defined more sharply in diagram $[U-W]$ vs $[B-U]$, so that not only the slope, but even the curvature of their mean path can be followed better. The best determined path is that of A 0, which occurs most frequently among the spectral type.

The most notable feature is the way in which the mean path of spectral type seems to follow the lines of constant temperature.

In Figure 8 one may note the enormous loop described by the A 0 type stars. The dashed part of this line is not based on observations. It is drawn only as a connection between the main-sequence stars and the supergiants. The latter drop sharply downwards to the left, i.e. to higher luminosities. Here we note again the tendency of the spectrum to approach a black-body distribution.

It is remarkable how the lines for spectral type B 8 through to A 2, seem to converge to one point.

However spectral type A 3–5 and later apparently are not able to converge to the same point. Less pronounced we see the same effect in Figure 7, where the mean path of all spectral types seemingly converge towards the same point, while type A 3–5 and F 0–2 diverge again.

8. The Emission-Line Stars

An extremely interesting group of stars in the Magellanic Clouds, are the emission-line stars. In Figure 7 they have been indicated as triangles. The bluest stars are the only two known Of stars in the Clouds. They fall nearly on the main-sequence and are practically unreddened. At $V-B=-0.02$ and -0.04 four stars of Wolf-Rayet type are shown. Still more towards the red follows a group of stars of spectral type B 2 or B 3. To this group a reddened star of type B 1 belongs, at $V-B=+0.04$. Above $V-B=0.05$ follows a group of seven stars, which are of spectral type B 8, B 9 or A 0. Among the latter two groups several of the stars are designated with eq by the Radcliffe astronomers, i.e. they are classified as P Cygni stars.

This remarkable array of emission-line stars of various type is especially significant in the case of the Magellanic Clouds. It indicates, that practically all stars with temperatures higher than 10000°, and which are the most luminous of their spectral group, are blowing matter into space.

In the colour-magnitude diagram, see Figure 2, these stars are all situated close to the lefthand part of the line of limiting bolometric magnitude. It is also clear from Figure 2, that the number of such stars, close to this line, is less in the Small Cloud than in the Large Cloud, which confirms the fact, that emission-line stars occur less frequently in the Small Cloud.

According to formulae (2) and (3) at a certain bolometric absolute magnitude g_{rad} surpasses g_{New} and more luminous stars therefore cannot exist. This computed limiting absolute bolometric magnitude is roughly one magnitude brighter than the line drawn

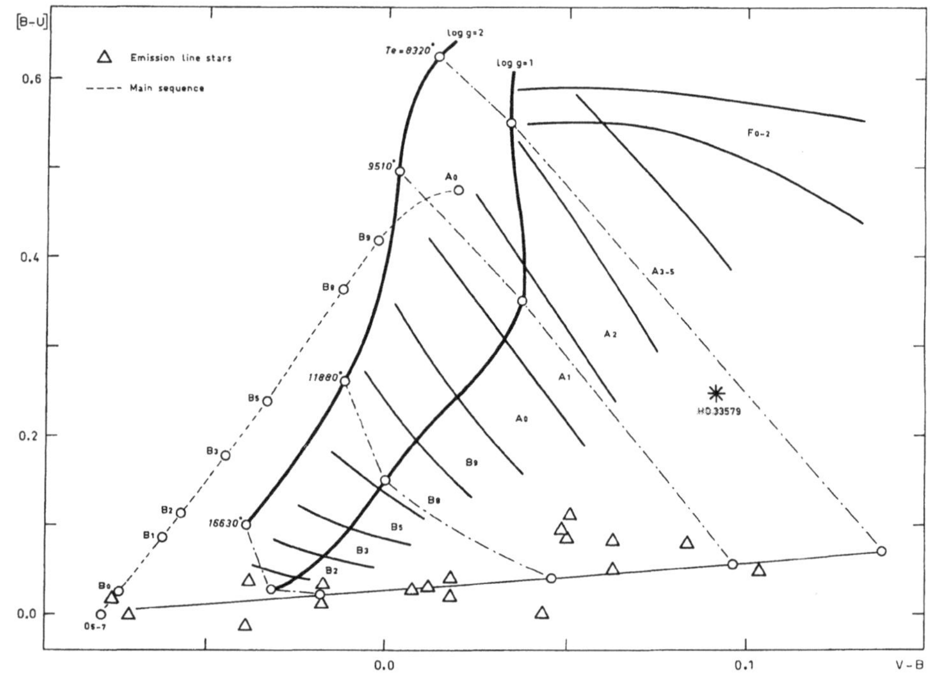

Fig. 7. Position of spectral types in colour-colour diagram. Main-sequence: circles on dashed line. Supergiants: boundaries between spectral-types are indicated as thin lines. Thick lines: (models) $\log g$ is constant. Triangles show emissionline and P Cygni type stars placed close to the black-body line.

in Figure 2, but in the estimate, turbulence was not taken into account. Altogether the evidence is strong, that in the Large Cloud several bright supergiants are at the limit of stability.

9. The P Cygni Stars

We draw attention to the fact, that all emission-line stars are situated close to the black-body line, see Figure 7. Judged from their brightness and spectral type alone, the P Cygni stars of spectral type B 9 should be placed somewhere near the line of $\log g = 1$, i.e. they have abnormal positions, which differ from the expected position on the blue envelope line, as if they are shifted to the right and downward roughly along a line of constant temperature. This strongly suggests, that the spectral distribution of these stars simulate stars of extremely low gravities. In other words, that not the emission-lines are responsable for the abnormal colours, but that the extremely low pressure in the atmosphere produces practically black-body colours, due to predominant electron scattering. This would explain, why we observe them on the black-body line. And extremely low pressure is to be expected, if the atmosphere is pushed outward by radiation.

It seems probable, that the P Cygni stars of type B 2 likewise are redder, than their normal counterparts by the same effect and not by interstellar reddening.

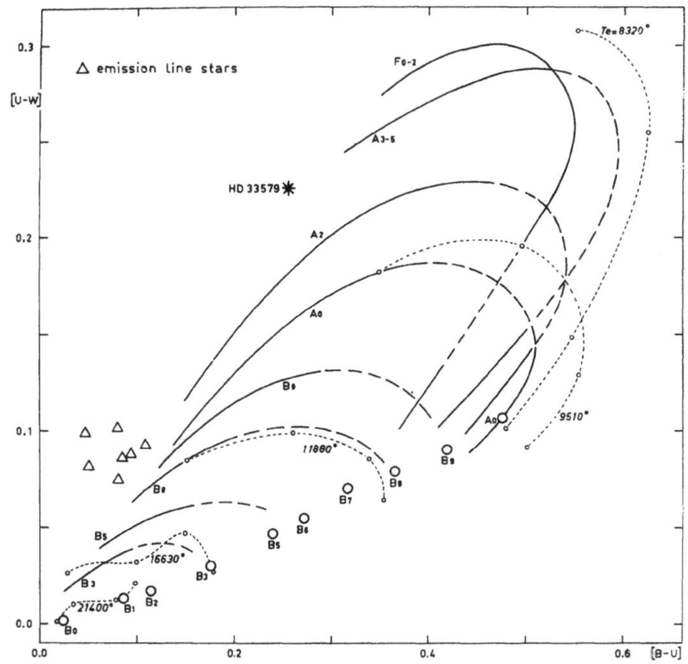

Fig. 8. Position of spectral type in colour-colour diagram $[U-W]$ vs $[B-U]$. Large circles: main-sequence. Small circles connected by thin dashed lines: constant effective temperatures. Drawn lines represent mean path of spectral type. Triangles: P Cygni type stars.

With respect to the Wolf-Rayet stars we are not certain about their intrinsic colours, because they occur mostly in heavily reddened regions surrounding the 30 Doradus nebula.

The phenomenon of the P Cygni stars affects our conclusion about the average interstellar reddening of the stars in the Magellanic Clouds. Many stars have been observed to the red side of the blue envelope lines. If their position were determined by interstellar reddening, the colour-excess derived from the different colour-colour diagrams should be the same. It turns out, that for more than half of the brighter stars this is not the case and the colour-excess derived from the diagram $[U-W]$ versus $V-B$ is considerable smaller than the colour-excess derived from the diagram $[B-U]$ versus $V-B$. This discrepancy is clearly seen for the brighter late *B*- and A-type stars.

On the other hand, when interpreted as a shift towards lower gravity, but at the same temperature, the discrepancy disappears in the same way as for the few stars which are known to be P Cygni stars.

There are also stars for which the assumption of interstellar reddening is valid. But, at least, for a large fraction of the stars, reddening must be interpreted not as due to interstellar dust, but as a change of intrinsic colour, produced by stirring up of the atmosphere, thus simulating a star of lower effective gravity.

Our observations suggest therefore, that the P Cygni phenomenon is in varying degrees of intensity present in all supergiant stars. In this connection it may be recalled, that practically all supergiants reveal emission in Hα at least.

A very important conclusion from the foregoing is, that the determination of luminosity from spectral criteria alone becomes very doubtful. A star may simulate, by the P Cygni effect, reduced gravity, which cannot be detected from the colours, the Balmer-jump or Balmer-lines. Perhaps a systematic observation, together with that of the normal criteria, of the intensity of Hα emission, may allow a correction for the effect. But at present no sufficient data are available to check this possibility.

We would like to conclude this paper with a short discussion of the parameters of the most luminous supergiant HD 33579. According to Feast *et al.* (1960) the spectrum is A3: Ia-0 (e). The emission refers to Hα and it is not the vigorous P Cygni phenomenon. The spectrum shows evidence of dilution of radiation in a very extensive atmosphere. Recently the spectrum has been analyzed by Przybylski (1968) and in Table IV some data are collected.

TABLE IV

A. Przybylski	*V, B, L, U, W*-photometry	
$V = 9.21$	$V = 9.14$	Wδ $= 1.40$
$B-V= 0.20$	$V-B = 0.092$	D $= 0.26$
$U-B= -0.27$	$[B-U] = 0.248$	$\theta = 0.580$
Wδ $= 1.40$	$[B-L] = 0.043$	$\log g_{\text{eff}} \simeq = 0$
D $= 0.30$	$[U-W] = 0.226$	
$\theta = 0.625$		
$\log P_e = 0$		

The colours of the *V, B, L, U, W*-photometry when transformed to UBV-photometry agree reasonably well. The strength of the hydrogen line as derived from $[B-L]$ agrees exactly with the value of Przybylski. The value of the Balmer-discontinuity D is somewhat different. However our numerous observations indicate that the Balmer-discontinuity is variable since the colour index $B-U$ varies with a total amplitude of about 0. 05. A large discrepancy is shown in the effective temperature which results from the use of a colour-spectrum relation by Przybylski, which is different from our relation. The extremely low electron pressure derived by Przybylski from the spectrum confirms the low value of $\log_{\text{eff}}$, that would follow from a rough extrapolation of the models.

From the data thus far collected, it would seem, that the star behaves normal and is exceptional only in the extremely low pressure of the atmosphere as follows from the various investigations. It therefore presents an extreme case of a quiet supergiant, not stirred up into strong P Cygni activity and we may state, that it dethrones α Cygni, which hitherto was considered as an extremely bright supergiant.

References

Feast, M. W., Thackeray, A. D., and Wesselink, A. J.: 1960, *Monthly Notices Roy. Astron. Soc.* **4**, 337.
Jager, C. de and Neven, L.: 1957, *Rech. Astron. Obs. Utrecht* **13**, No. 4.
Jager, C. de and Neven, L.: 1967, *Bull. Astron. Inst. Neth. Suppl.* **2**, 125.
Mihalas, D.: 1966, *Astrophys. J. Suppl.* **114**, 1.
Przybylski, A.: 1968, *Monthly Notices Roy. Astron. Soc.* **139**, 313.
Underhill, A. B.: 1968, *Astr. Astr. Phys.* **6**, 39.
Walraven, Th. and J. H.: *Bull. Astron. Inst. Neth.* **XV**, 496.

CEPHEID VARIABLES AND THE MAGELLANIC CLOUDS

R. F. CHRISTY
California Institute of Technology, Pasadena, Calif., U.S.A.

1. Introduction

Because the Magellanic Clouds provide a large number of variables at approximately the same distance and with little differential reddening, they provide a unique laboratory for the study of cepheid variability. Much more systematic observation of the variables will be necessary to fully exploit the possibilities they provide. It is already clear that the Magellanic Cloud variables differ from the galaxy variables in the number of variables at different periods, in the period amplitude relation, and, apparently, in the mean color. These differences must be due to intrinsic differences in the nature of the stellar populations of the Clouds and of the Galaxy. Finally, it has been found that certain non-linear features of cepheid variables can be correlated with stellar radius and luminosity. In the case of the Clouds, these features can be used to deduce distances in a new way.

2. The Instability Strip

Systematic calculation (1) of pulsation instability for a wide variety of stellar models has shown that the stability of a star is determined by T_e, surface g, and hydrogen-helium ratio. The instability moves to lower T_e for smaller g and for smaller helium content. Since the period of the fundamental mode, P_F, is principally dependent on g, it is understandable that we can also define an instability strip in the $\log P_F - \log T_e$ plane. For Y (helium mass fraction) equal to 0.3, we find the high T_e boundary of pulsation instability is given by

$$\log T_e = 3.828 - 0.06 \log P_F \text{ (days)}. \tag{1}$$

We can include the effect of helium by the relation

$$\log T_e = 3.798 + 0.1\, Y - 0.06 \log P_F \text{ (days)}. \tag{2}$$

The low T_e boundary of instability is estimated to lie about 0.06 lower in $\log T_e$.

Using the relation (1) above, and the calculated

$$P_{F\,\text{days}} = 0.0224 \frac{(R/R_\odot)^{1.74}}{(M/M_\odot)^{0.7}}, \tag{3}$$

we can derive a theoretical Period-Mass-Luminosity relation following the center line or the boundaries of the instability region. In this way we get a relation that is not dependent on limited stellar statistics but is, in general, difficult to compare with

Muller (ed.), The Magellanic Clouds, 136–143.

observation. It is

$$-M_b = 2.01 \log M/M_\odot + 2.28 \log P_F + 0.37 \quad (4)$$

for the center of the strip for $Y=0.3$. This relation reduces to a Period-Luminosity relation when the Mass-Luminosity relation is introduced.

For population II variables, we assume $M/M_\odot = 0.5$ and get

$$-M_b = -0.23 + 2.28 \log P_F. \quad (5)$$

For population I variables, if $L \propto M^n$ we get $-\Delta M_b = A \log P_F$ where $A = 2.28/(1-(0.805/n))$ and for $n=3.5$, $A=2.96$ which is close to the relation of Gascoigne and Kron [2].

3. Comparison with Observation

Gascoigne and Kron have found that the cloud cepheids are bluer than in the galaxy by as much as 0.2 in $B-V$ at the shorter periods. I will assume that their conclusion is

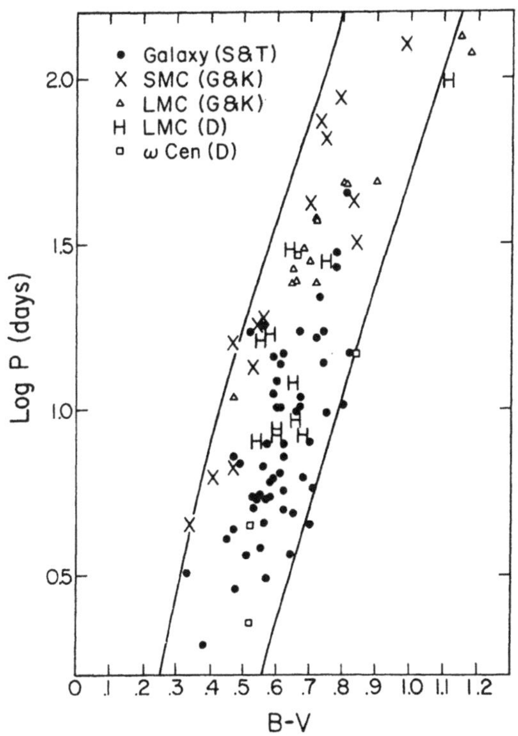

Fig. 1. Period-color diagram for cepheid-type variables. Data sources: Galaxy – Sandage and Tamman [4]; SMC and LMC – Gascoigne and Kron [2]; LMC and ω Cen (Dickens, R. J. and Carey, *Roy. Obs. Bull.* No. 129). The solid lines mark the calculated blue boundary and estimated red boundary of instability.

correct and will discuss how we may interpret it. I will use the relation

$$\log T_e = 3.886 - 0.175\langle B - V\rangle \qquad (6)$$

from Kraft [3] where necessary to translate from $\log T_e$ to $\langle B-V\rangle$. Actually, this relation should also depend on g and therefore should be different for the short and long period cepheids. Such a refinement will have to await new work on calibration of temperatures of stellar atmospheres.

First, all indications are that the helium content of the clouds differs little from that near the sun. Now it would take an increase of Y by 0.3 to shift the instability strip by 0.2 in $B-V$ toward the blue so I will assume that the instability strip in the clouds is essentially identical in color with that of the Galaxy.

In Figure 1, I have plotted the mean unreddened colors of observed cepheids in a color-period plane. We then find that the observed variables lie inside a strip of width ≈ 0.37 in $\langle B-V\rangle$ but becoming narrower (≈ 0.2 in $B-V$) for RR Lyrae variables. Except for two or three rare examples (such as Y Oph), the classical cepheids of our Galaxy [4] occupy a strip of width only ≈ 0.2 in $B-V$ and are confined to the red side of the wider instability region. The observations of Gascoigne and

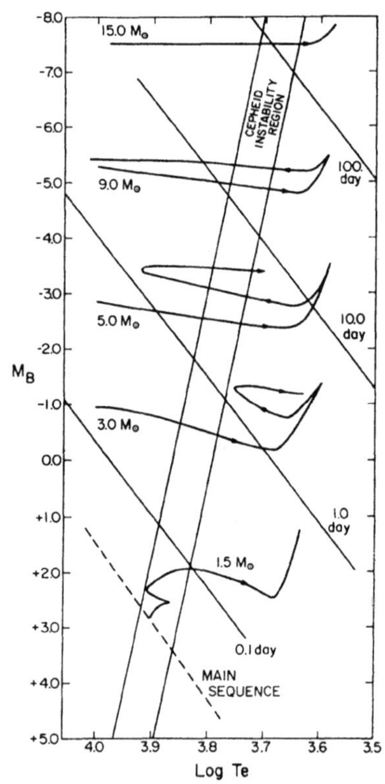

Fig. 2. Evolutionary tracks after Iben [8]. Constant period lines for fundamental mode vibration are shown approximately.

Kron [2] and of Dickens [5] in the Magellanic Clouds are primarily to be found on the blue side of this instability region. It is also well-known [6, 7] that there is a high frequency of short period (1–2 days) cepheids in the Small Magellanic Cloud, whereas such variables are almost absent our in Galaxy.

We propose the following interpretation of the observed mean colors and frequency of occurrence of various periods (see Figure 2). The evolutionary calculations of Iben [8] and of Hoffmeister *et al.* [9] show a first crossing of the cepheid instability strip from left to right on the way to the first red giant stage. However, a much slower crossing, during core helium burning, occurs after the first red giant stage when the evolutionary track makes an excursion to higher T_e and returns, after a loop, to the red giant stage. These calculations show that the loop extends further to the blue the higher the luminosity. In this way, these authors have explained a lack of short period cepheids, when the loop does not intersect the instability region, followed, at higher luminosity, by a peak in the frequency when the loop just reaches the strip and then returns. At higher luminosity, the crossing of the strip is complete and is more rapid, corresponding to a lower frequency of cepheids.

It was noted by Hoffmeister [10] that the loops in the evolutionary track, which are so essential to the cepheid phase, are prominent for a set of calculations with $Z=0.044$ and $Y=0.354$ but almost disappear for a set of tracks with $Z=0.021$ and $Y=0.24$.

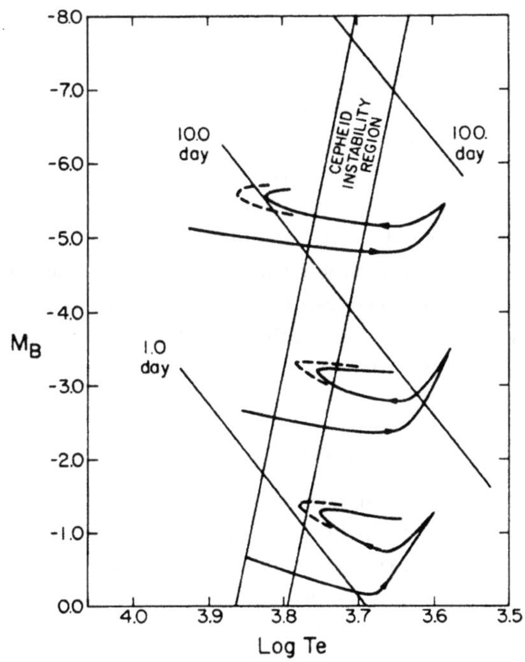

Fig. 3. Schematic evolutionary tracks to illustrate how a proposed difference between galactic evolution (solid lines) and SMC evolution (dashed lines) might explain the differences in the cepheid variables.

This implies that the frequency of cepheids in a given population is very sensitive to either Z or Y or both. I now propose that the cepheids in our Galaxy differ from those in the SMC by a small amount in Z or Y so that the evolutionary tracks show somewhat shorter loops for our Galaxy. This view is illustrated schematically in Figure 3. We would then understand the excess of short period cepheids in the SMC as due to the greater extent to the blue of the evolutionary tracks at low luminosity compared to the Galaxy. At the same time, these tracks would populate the blue side of the instability strip whereas in the Galaxy, when the tracks do intersect the instability strip it would be primarily on the red side and we would thus understand the absence of cepheids, in the Galaxy, on the blue side of the strip. Some of these same differences appear to exist, in a milder form, between cepheids in the direction of the galactic center and these away from the center which are more like those in the Magellanic Clouds. We thus see that the frequency of short period cepheids is a very sensitive indicator of some, not yet fully understood, aspect of abundances. Further calculation of evolutionary models is needed to clarify the meaning of these evolutionary tracks.

4. Radii from the Hertzsprung Progression

One of the most characteristic features of cepheid light curves in the period range 7–15 days is the appearance of a second bump in the light curve at different phases, closely correlated with the period (Figures 4, 5). Recent model calculations have reproduced the interior motions of variables which show this phenomenon (Figure 6). These motions show that the second bump phenomenon is associated with compressional waves, originating near minimum radius in the helium ionization layer, which travel to the center and out again, arriving at the outside some 1 to 1.5 periods later. These waves appear to reinforce by constructive interference just in those cepheid variables of 7–15 days period where the phenomenon is observed. For periods longer than 12–15 days, this reflected wave remains strong but arrives back nearly one period later and serves to reinforce the basic pulsation. This mechanism thus explains the increase in amplitude for variables of period greater than 15 days.

A systematic study of models in which this reflected wave is prominent showed that the delay of the reflected wave after the primary is correlated with the mean radius of the model according to

$$R/R_{\odot} = 4.05 \times \text{delay (days)}. \tag{7}$$

In this expression the times are measured from the center of the primary outward acceleration phase to the center of the secondary outward acceleration phase.

We can combine with this expression, the period-mass-radius relation (3) so that the observed period, combined with an observed second bump gives a value for the mass. Masses determined in this way are typically about 60% of those determined for the same luminosity from evolutionary calculations. We can also combine the radius determined in this manner with a value for T_e determined from the mean color [6] to arrive at the mean absolute magnitude of the variable.

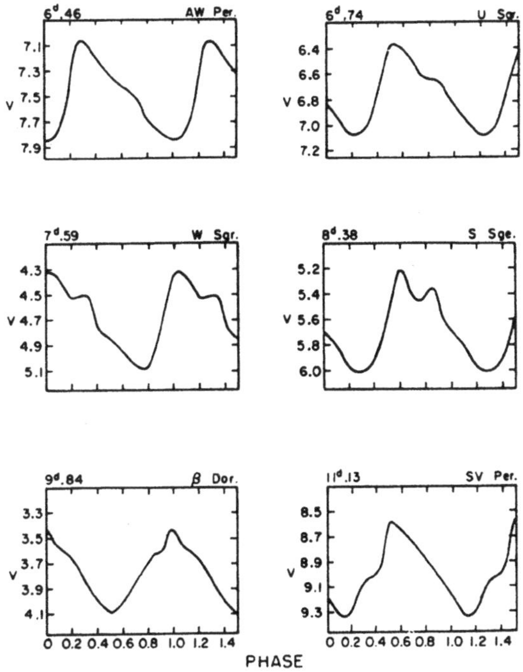

Fig. 4. Observed luminosity variation for six cepheid variables.

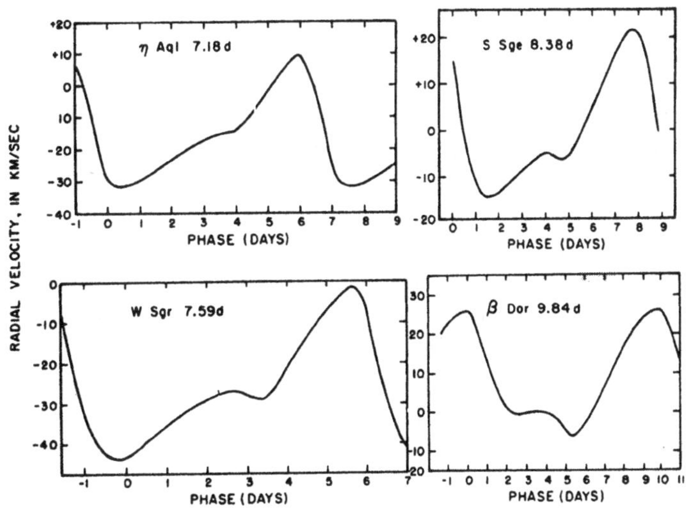

Fig. 5. Observed radial velocity variation in four cepheid variables: η Agl and W Sgr from Jacobsen, T. S.: 1926, *Lick Observatory Bull.* **12**, 138; β Dor from Applegate, D.: 1927, *Lick Observatory Bull.* **13**, 12; S Sge from Herbig, G. H. and Moore, J. H.: 1952, *Astrophys. J.* **116**, 348.

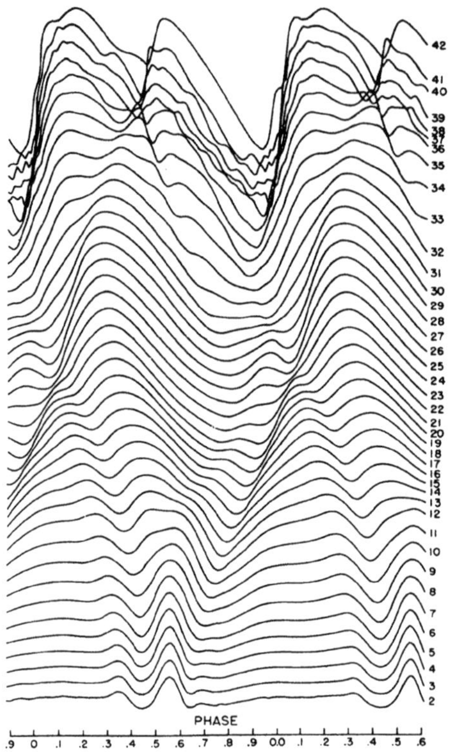

Fig. 6. Calculated velocities of all mass zones at maximum amplitude for a model simulating β Dor. The zeros are progressively shifted and the scales progressively enhanced toward small radii (zone 2) in order to make visible the pattern of in and outgoing signals. Velocities outward from the star are plotted upward. The topmost curves refer to optical depths less than unity. Zone 2 is at $r/R \approx 0.1$.

This procedure has so far been applied to only a very few variables. S Nor [11] is perhaps the best example of a cluster variable for calibration purposes. I have estimated the delay of the reflected wave as $1.33 \times 9.75 = 12.96$ days so $R/R_\odot = 52.5$. Taking $\langle B-V \rangle = 0.75$, we get $\log T_e = 3.755$, $\log L/L_\odot = 3.408$, and $M_b = -3.80$, whereas Kraft's [3] value is -4.05. This serves to emphasize the fact that the relation [7] leads to consistently smaller radii than is customary and to correspondingly smaller masses.

The best case for application of this method in the clouds appears to be HV 2432 ($P = 10.92$ d) [2] in the LMC. It shows a prominent second bump in the light curve and, by comparison with S Sge which is similar and where a velocity curve is known, we deduce a delay of 1.43 periods and $R/R_\odot = 63.2$. Models of these characteristics have $Q = 0.0435$ and so we get $M/M_\odot = 4.0$. The mean $\langle B-V \rangle$ of this variable is 0.52 and, corrected for a reddening of 0.05 gives $\langle B-V \rangle_0 = 0.47$. Then from (6) we get $\log T_e = 3.804$ and $\log L/L_\odot = 3.766$ and $M_b = -4.695$. Now $V = 14.23$ so $V_0 = 14.08$ and $m_b = 14.06$. The modulus is thus 18.75.

No variable measured by Gascoigne and Kron in the SMC shows a similar prominent bump but variables 1400 ($P=6.65$ d) and 827 ($P=13.47$ d) closely bracket the range where a prominent bump is anticipated and the distance modulus of the SMC is determined to be 0.4 greater than for the LMC by the requirement that neither of these variables should show the bump.

These moduli are greater by 0.25 than those of Sandage and Tamman. The increase over an earlier report is solely due to an increase in $\log T_e$ from a change in (6) by 0.025. This immediately results in an increase in brightness by 0.25 and also in modulus. Alternatively, we can use this method as a device to relate galactic and cloud variables without resorting to absolute calibration. If we adopt Kraft's calibration of S Nor, we would then deduce a modulus of 19.0 for the LMC and 19.4 for SMC. This tentative conclusion emphasizes the need for more careful study of galactic cluster variables and cloud variables which show this bump.

I have made a preliminary exploration of possible sources of error in the above radius and mass determinations. If the Cox and Stewart opacities are consistently too low by a factor of two in the temperature range from 50000 K to 500000 K, then the radii could be increased by 12% and the masses by 0.4%. Such a change would, however, not change the direct comparison of the LMC and S Nor.

Acknowledgements

This research was supported in part by the National Aeronautics and Space Administration grant NGR 05-005-007, by the National Science Foundation grants 9344 (formerly 7976) and 9114, and by the Office of Naval Research contract Nonr-220(46).

References

[1] Christy, R. F.: 1968, *Q. J. Roy. Astron. Soc.* **9**, 13; 1966, *Ann. Rev. Astron. Astrophys.* **4**, 353; 1968, unpublished calculations.
[2] Gascoigne, S. C. B. and Kron, G. E.: 1965, *Monthly Notices Roy. Astron. Soc.* **130**, 333.
[3] Kraft, R. P.: 1961, *Astrophys. J.* **134**, 616.
[4] Sandage, A. and Tammann, G. A.: 1968, *Astrophys. J.* **151**, 531.
[5] Dickens, R. J.: 1966, *Observatory* **86**, 18.
[6] Arp, H. C.: 1960, *Astron. J.* **65**, 404.
[7] Payne-Gaposchkin, C. and Gaposchkin, S.: 1966, *Smithsonian Contrib.* **9**.
[8] Iben, I., Jr.: 1967, *Ann. Rev. Astron. Astrophys.* **5**, 571.
[9] Hoffmeister, E., Kippenhahn, R., and Wiegert, A.: 1964, *Z. Astrophys.* **60**, 57.
[10] Hoffmeister, E.: 1967, *Z. Astrophys.* **65**, 164.
[11] Feast, M. W.: 1967, *Monthly Notices Roy. Astron. Soc.* **136**, 141.

EVOLUTION OF MASSIVE STARS

R. KIPPENHAHN

Universitäts-Sternwarte, Göttingen, W. Germany

In the past, the work in stellar evolution theory dealt mainly with stars in the range between 1 and 10 solar masses. As a result of this one has obtained a fairly good understanding of the main sequence turnoff points, of the Hertzsprung gap and of the helium burning red giants, including their cepheid stages. Unfortunately, the theoretical work which has been done for stars above 10 solar masses suffers from uncertainties in the theory of stellar structure, especially in the stability theory which decides what part of a stellar model will be convective and which not. Up to now there is no general agreement on whether or not the supergiants in young clusters like $h+\chi$ Per (Figure 6) in our galaxy or NGC 2004 (Figure 4) in the LMC have already finished their central helium burning and are in stages of higher nuclear burning.

1. The Upper End of the Main Sequence

It is commonly known that the main sequence life time τ_{MS} decreases with the mass M of a star, since the luminosity goes roughly like M^3 while the nuclear fuel available is only proportional to M. But, the more massive a star the flatter the mass luminosity relation and for the limit case of extremely massive stars one expects $L \sim M$. Therefore, the decrease of τ_{MS} with increasing mass will become less pronounced for massive stars (Figure 1). This effect is due to the influence of radiation pressure which for massive stars becomes more and more important.

Radiation pressure also influences the vibrational stability of massive MS stars. The more the radiation pressure becomes important in the equation of state, the more the star's central regions participate in an eventual radial pulsation of the star. For stars of medium mass the amplitudes of radial pulsation drop rapidly if one goes inwards and therefore the region of nuclear energy generation does not participate in any pulsation. As a consequence the nuclear reactions, which in principle could drive pulsation, have no chance to overcome the damping which takes place in the outer layers and which is favoured there by the large amplitudes of radial pulsation in the outer layers. The situation is different for massive stars where the central region participates much more in the radial pulsation such that, during each phase of maximum contraction, the nuclear energy output is increased and energy is used to push the outer layers outwards again (ε-mechanism). Stars above about 65 solar masses should be vibrationally unstable on the main sequence. Up to now it is not yet known what the consequence of this instability will be, since the type of nonlinear investigation, which has been carried out for cepheids by Christy, by J. Cox and his collaborators and most recently by Baker and v. Sengbusch with great success, has not yet been done for the massive unstable main sequence stars. Will the stars just

Muller (ed.), The Magellanic Clouds, 144–155.

show some irregularities in their luminosity? Or will they pulsate with growing amplitudes until during each period the outer layer reaches the escape velocity? In the latter case they may shed mass during each cycle of pulsation until their total mass has become sufficiently small so that the star settles down as a stable star, below the mass critical for the vibrational stability behaviour. One might ask whether the observed emission lines in the spectra of the brightest stars in the MC's are connected with the vibrational instability of hydrogen burning stars which are too massive to be stable.

From the arguments given above we would expect the following picture of the evolution of a cluster containing massive stars: After the stars have settled down on

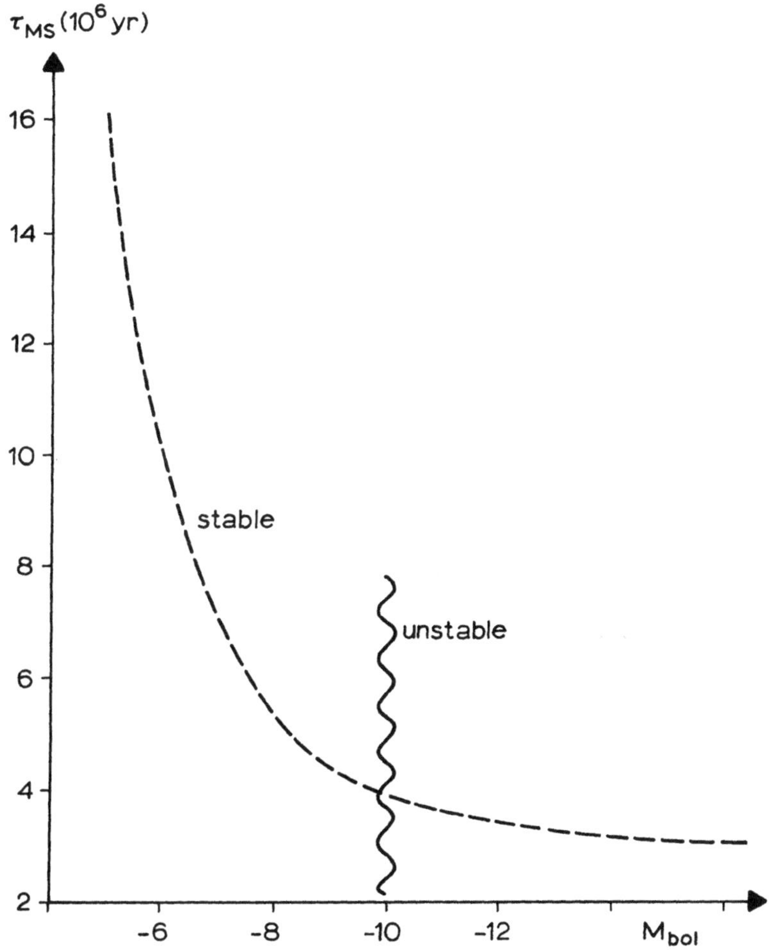

Fig. 1. The main sequence lifetime τ_{MS} of (population I) stars of different main sequence luminosity (schematically). For the most massive stars τ_{MS} becomes gradually constant. Main sequence stars brighter than $M_{\mathrm{bol}} = -10$ should be vibrationally unstable.

the main sequence, those with $M_{bol} < -10$ will immediately become vibrationally unstable. Then, after a main sequence life time of about 4×10^6 yr, the stars with $M_{bol} < -8$ will exhaust their central nuclear fuel and leave the main sequence almost simultaneously. Afterwards, the main sequence stars of lower luminosity will gradually finish their central hydrogen burning, causing the turnoff point to go to fainter and fainter stars.

For the upper end of the MS where the vibrational instability occurs, the bolometric corrections may be roughly of the order 3^m (Figures 2a, b). If we are looking for unstable stars in the MC's we therefore should look for MS stars of $(B-V)_0 \approx 0.25$ and $M_v < -7$ or $m_v < 12$.

2. The Clusters of Medium Age

If one tries to compare theoretical evolutionary tracks in the theoretical Hertzsprung-Russell diagram with log T_{eff} and M_{bol} as ordinate and abscissa with observed color-magnitude diagrams, one has to take into account the rather difficult transformation which is involved. A rough sketch of this transformation – based on new results by Wilde (1969) – is given in Figures 2a, b. (I am glad that these figures gave me a chance to honour that part of the world which is sponsoring this conference so generously.)

One of the most interesting clusters for comparison of theory with observation is NGC 1866 in the LMC. It contains a well defined upper end of a main sequence and a group of red giants. Among the evolved stars are 13 cepheids. Arp (1967) has already tried to compare the HRD of the cluster with evolutionary tracks. A more detailed comparison has recently been made by Meyer-Hofmeister (1969). She computed evolutionary tracks in the range between 3.5 and 5.5 solar masses, constructed isochrones and tried to compare position and frequencies in the distribution of the observed stars with her theoretical models. In addition she investigated the vibrational instability of the models in the cepheid strip in order to compare the periods of the cepheids observed in the cluster and the periods theoretically obtained for the most unstable models. She was able to get all the main features observed (Figure 3) and to obtain roughly the right periods. One can conclude, therefore, that we understand the evolution of the stars in this cluster.

Figure 4 gives a composite HRD of clusters in the LMC (Hodge, 1966). One can guess that clusters having a turnoff point below 10 solar masses ($m_v \geqslant 16$) are with respect to their evolution similar to NGC 1866, the red giants mainly being in the stage of central helium burning.

For lower luminosities one already gets red giant branches, which belong to clusters which have HRD's like the globular clusters of our galaxy. Since up to now there is no complete theory of horizontal branch stars, we are not yet able to give good models for the stars in these clusters and compare theory with observation. But there is no doubt, that all the red giant stars on the upper end of the ascending branch of the population II clusters, and also the horizontal branch stars are in the stage of central helium burning.

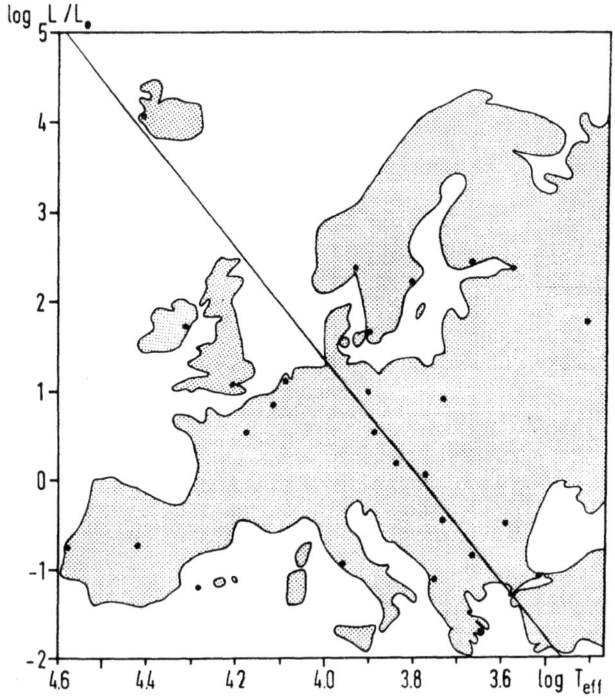

Fig. 2a.

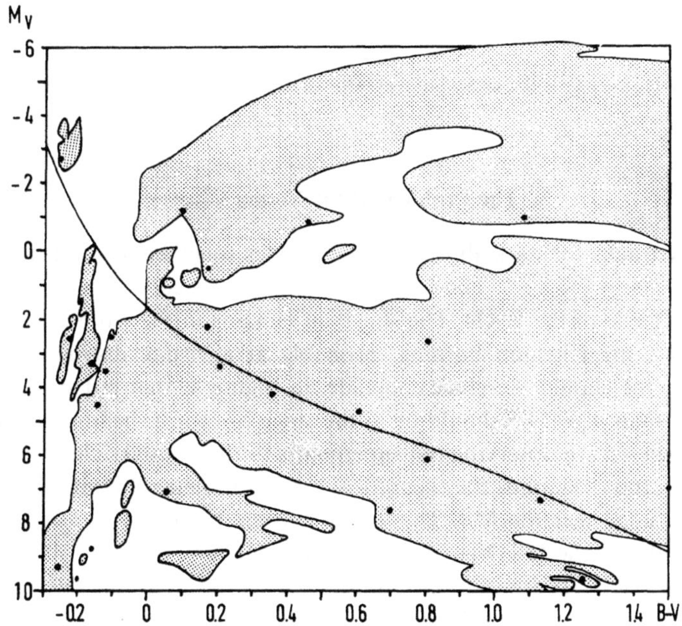

Figs. 2a–b. The transformation of the theoretician's HRD above into that of the observer. The straight line on Figure 2a gives the theoretical main sequence which takes on the well known bent form when it is transformed into the observers diagram. The transformation has been derived by Wilde (1969).

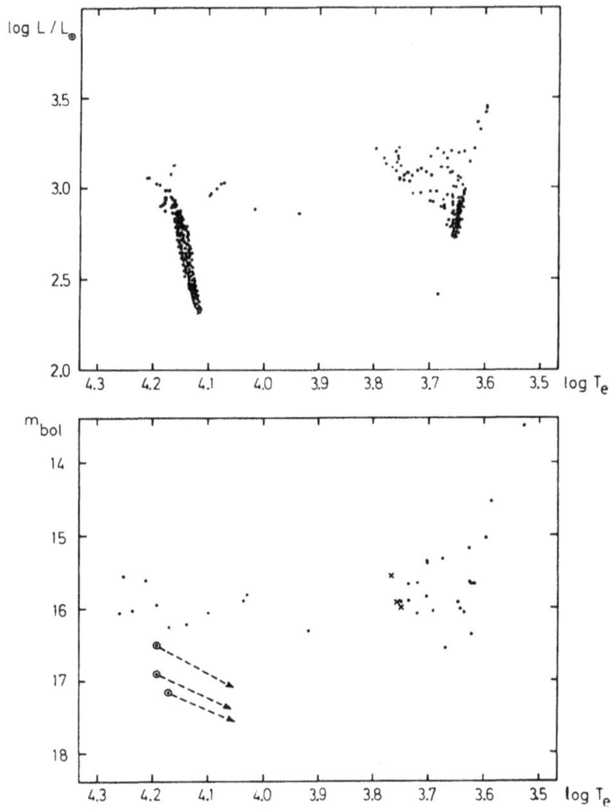

Fig. 3. Meyer-Hofmeister's theoretical model (above) for the observed HRD of NCG 1866 (below).

3. The Problem of the Young Clusters

Evolutionary tracks for stars above 10 solar masses together with a track for 9 solar masses are given in Figure 5. The star of 9 $M_{\odot}$ ignites its central helium while it is on the right hand side of the HRD. The star then remains in the red giant region during the phase of central helium burning, probably at a visual absolute magnitude of about -4.5. One should therefore conclude, in Figure 4, that red supergiants in the LMC fainter than $m_v = 14.5$ should be in the stage of central helium burning.

But although we are observing red supergiants in the younger clusters of the LMC like in NGC 2100 (Figure 4) the theoretical models in general ignite their helium on the left hand side, near the MS. The young clusters therefore give rise to the same kind of difficulty in the HRD as h + χ Per in our Galaxy. Figure 6 shows this group of stars in the HRD. Tedre is a well pronounced clustering in the red giant region at $(B-V)_0 \approx 1.8$ similar to that in NGC 1866 (Figure 3); it is well isolated from the main sequence. But while the theoretical models for NGC 1866 demand this type of

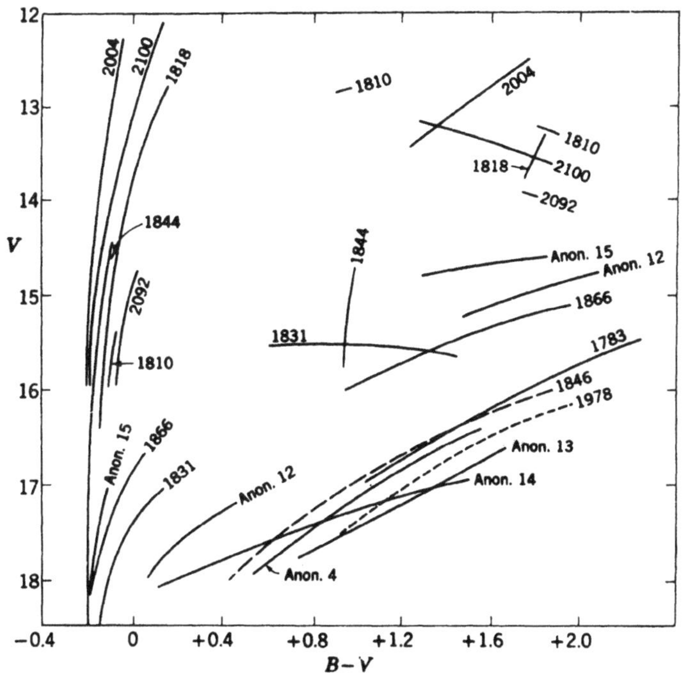

Fig. 4. Composite HRD for clusters in the LMG (after Hodge, 1966).

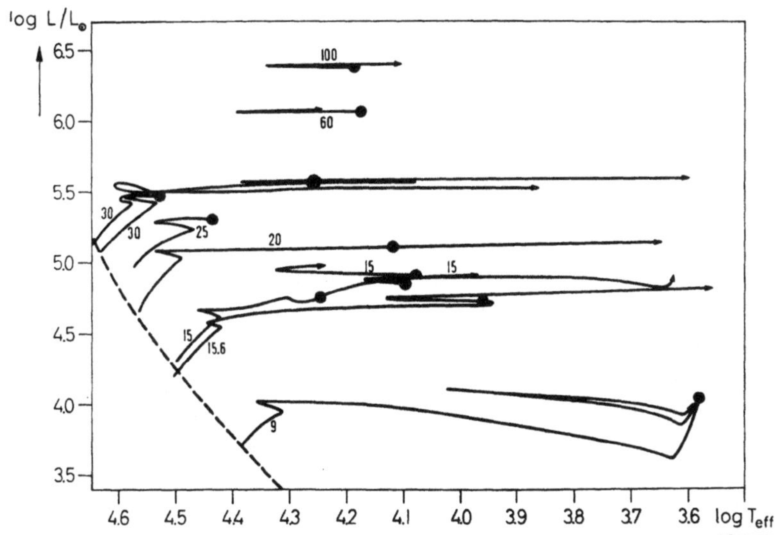

Fig. 5. Theoretical evolutionary tracks for different masses. The dots indicate the ignition of central helium. The numbers give the masses in solar units. The different tracks are computed by Hofmeister (9 $M_\odot$, 1967); Hayashi and Cameron (15.6 $M_\odot$, 1962); Iben (15 $M_\odot$, 1966); Kotok (20, 30 $M_\odot$, 1965, 1966); Kippenhahn (25 $M_\odot$, 1969); Stothers (30 $M_\odot$, 1966); Stothers-Chin (15, 60, 100 $M_\odot$, 1969).

clustering, central helium burning seems not to be responsible for it in the case of younger clusters where more massive stars are involved.

Two possible explanations for this problem have been given. In the first one it is assumed that helium burning indeed takes place on the left hand side near the main sequence as suggested by the theoretical evolutionary tracks. After exhaustion of

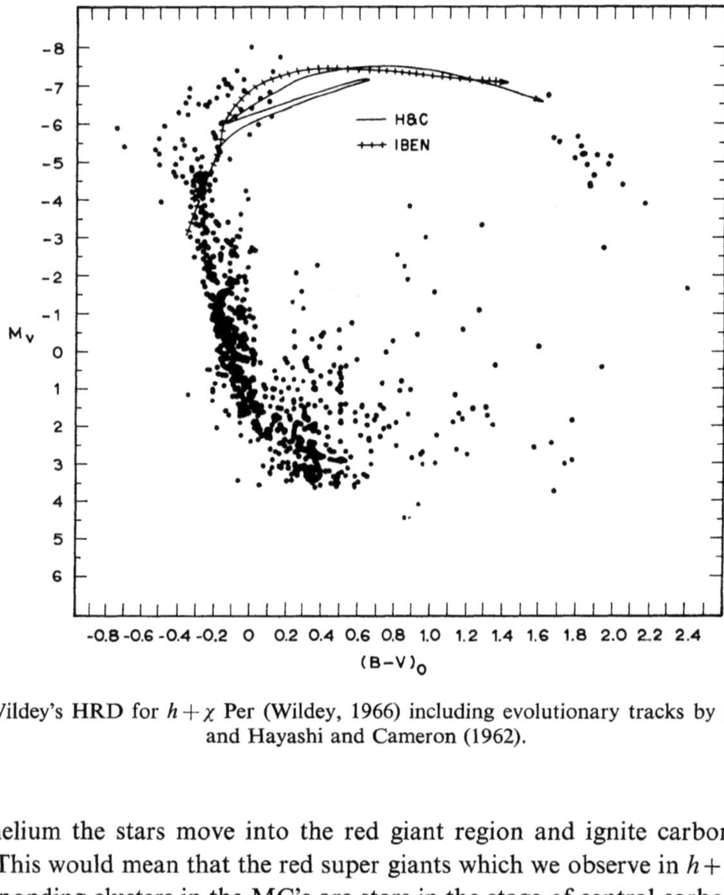

Fig. 6. Wildey's HRD for $h+\chi$ Per (Wildey, 1966) including evolutionary tracks by Iben (1966) and Hayashi and Cameron (1962).

central helium the stars move into the red giant region and ignite carbon in their centers. This would mean that the red super giants which we observe in $h+\chi$ Per and in corresponding clusters in the MC's are stars in the stage of central carbon burning and higher nuclear reactions. This explanation is grounded on an unquestioning faith in evolutionary calculations. A rather lengthy discussion of this way of getting around the difficulty has been given in literature during the last years (Hayashi and Cameron, 1962, 1964, Wildey, 1966). Up to now no really good models have been computed from the MS through helium and carbon burning. Estimates indicate that the time of carbon burning for a 15.6 solar mass star could be on the order of 2.3×10^5 yr, a number which is reduced by more than half if neutrino losses are taken into account. This makes it difficult to explain the large number of stars in $h+\chi$ Per which appear to be in this stage rightnow. I personally think, that this time estimate should be checked by better model calculations. But since there are good physical reasons to believe in the neutrino losses, one might try to find another explanation for the red supergiants in young clusters.

Another possible explanation assumes that in spite of the fact that most of the model calculations indicate the contrary, the red supergiants in young clusters are burning central helium. Indeed, model calculations become more questionable if one goes to fifteen or more solar masses. In fact, at these masses, it is fairly easy to construct stellar models in the stage of central helium burning which are in the right place in the HRD.

The easiest way of finding such models is the concept of *generalized main sequences* (Kippenhahn and Weigert, 1967) as it has been applied in order to understand models after mass exchange in close binary systems. For this purpose one constructs models consisting of a helium core and a hydrogen rich envelope. The models have a discontinuous transition from the helium core to the hydrogen rich mixture at a certain depth defined by $M_r/M=q_0$. If q_0 is given, a stellar model can be constructed for each mass M which obtains all its luminosity from nuclear burning, which according to the value of q_0 can be either hydrogen or helium burning, or both, taking place in the same model – the helium burning in the center and the hydrogen burning in a shell at q_0. $q_0=0$ gives the normal main sequence for hydrogen burning (Figure 7). $q_0=1$ gives the helium main sequence. The other values of q_0 give other 'main sequences' of a more general form. They cover a large part of the HRD, some regions are covered twice.

From the generalized main sequences one can conclude that there are stellar models in the upper right corner of Figure 7 ($\log T_{\text{eff}} \approx 3.6$, $\log L/L_\odot \approx 4.5$). Probably the generalized MS for $q_0 \approx 0.5$ would give the best models for the red supergiants in young clusters, i.e. that stars of about 5 solar masses, with helium burning in a core of 2.5 solar masses and hydrogen burning in a shell at the surface of the helium core, would give just the right red supergiants. One would not have to worry about the time scales, since the stars on the generalized main sequences live on a long nuclear time scale, given in our case by the central helium burning.

Thus, it is not difficult to invent stars which could represent the red supergiants; the difficulty is to explain how models like our artificial ones can be formed in actual evolution. We know from the turnoff points that the masses of the evolved stars should be at about 15 solar masses. But, since on our generalized main sequences at the place where the stars of interest are, the masses are of the order of 5 solar masses, one could assume, for instance, that they originally had been stars of 15 solar masses which, during hydrogen burning, have formed helium cores of about 3 solar masses. The stars must then have undergone mass loss whereby they end up as helium burning models in the red giant region as stars of 5–6 solar masses. We know from the generalized main sequences that they then would be just in the right place.

I do not think that the assumption of mass loss is really necessary. Unfortunately, the generalized main sequences have not been computed for masses higher than 5 solar masses, but it is well known that the luminosity of a star with a helium core and with hydrogen shell burning remains roughly the same if the mass contained in the hydrogen rich envelope is changed. This is known from the computation of mass exchange in close binaries where similar models, after mass loss, resume the same luminosity they had before they lost most of the mass of their hydrogen rich envelope. We therefore can expect, that stellar models of 15 solar masses with helium cores of 3 solar masses would be in the right position in the HRD.

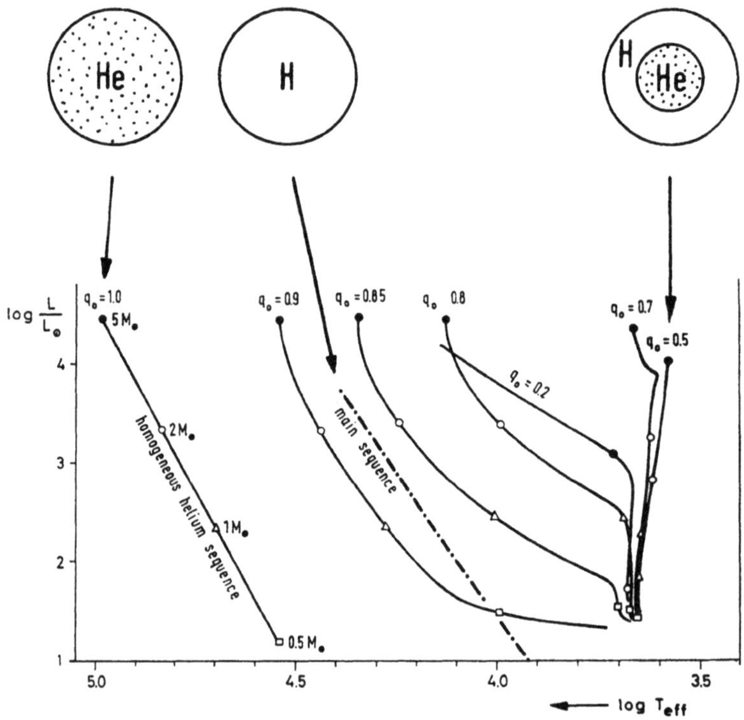

Fig. 7. Generalized main sequences covering the HRD. The disks on top indicate schematically on a mass scale the internal chemical structure of the models which belong to the parameters $q_0 = 0$, 0.5, 1 (Giannone, *et al.*, 1968). The symbols ●, ○, △, □ indicate 5, 2, 1, 0.5 $M_\odot$.

The only difficulty is that, for the generalized MS of Figure 7, it is assumed that there is a discontinuous transition in chemical composition from the core to the hydrogen rich mixture of the envelope.

It is difficult to see how, during evolution on the MS and afterwards, such a discontinuity could be formed. In all cases which have been computed, it turns out that a fairly broad transition region is formed. In Figures 8 and 9, some results are summarized which I recently obtained for a star of 25 solar masses. As a stability condition, which determines the boundaries of the convective layers and which therefore rules the distribution of the helium produced during MS burning, I took the Schwarzschild criterium. As one can see from the lower part of Figure 9, there are broad smooth transition regions while the region of varying hydrogen content shows only a small discontinuity. Actually the region from $M_r/M=0.3$ to $M_r/M=0.6$ can be considered as a broad region of variable chemical composition. Due to this fairly broad transition region, helium burning takes place near the MS (see 25 solar mass track in Figure 5). The mixing would have been less, if I would have used the Ledoux criterium which takes into account the stabilizing effect of the gradient of the molecular weight.

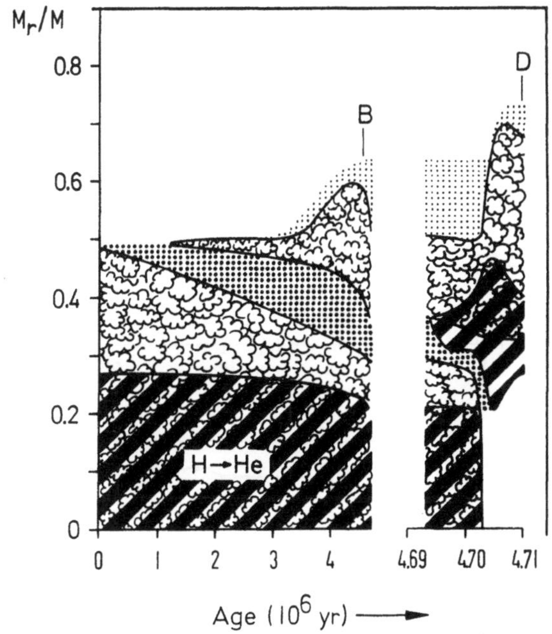

Fig. 8. Variations in time of the interior of a star of 25 $M_\odot$ during exhaustion of central hydrogen. The abscissa gives the age in 10^6 yr, the ordinate is M_r/M. 'Cloudy' regions indicate convection. The stripes indicate regions in which the nuclear energy generation exceeds 10^3 erg/g sec. Heavy dots indicate regions where the molecular weight increases inwards but which are vibrationally stable ($\nabla < \nabla_{ad}$). Light dots are plotted in layers where the molecular weight increases inwards but which are vibrationally marginal stable ($\nabla = \nabla_{ad}$).

I would then probably have obtained models which ignite their helium a bit farther to the right. The steeper the transition from helium to hydrogen rich outer layers, the more one approaches the models having a discontinuous transition, which, according to the generalized MS, should burn their helium in the red supergiant region. Indeed, authors who have less mixing and therefore steeper transition zones obtain helium burning stars farther in the red giant region of the HRD. A typical example is the model of 20 solar masses computed by Kotok (1966). As one can see from Figure 5, his models ignite helium at about log $T_{eff} \approx 4.1$. Then the star moves towards the right where it slows down and uses up its central helium within 10^6 yr, one eighth of its main sequence lifetime. Similar arguments favouring a suppression of mixing in order to get the models into the red giant region have been given by Paczynski (1967). The difficulty is that all the evolutionary tracks for massive stars suffer from the lack of a good theory of mixing. There is a great uncertainty as how to deal with the so-called semi-convective regions. It is in these zones of semiconvection that the chemical transition layers are formed. For massive stars, rotation may also play a role in the formation of chemical transition layers.

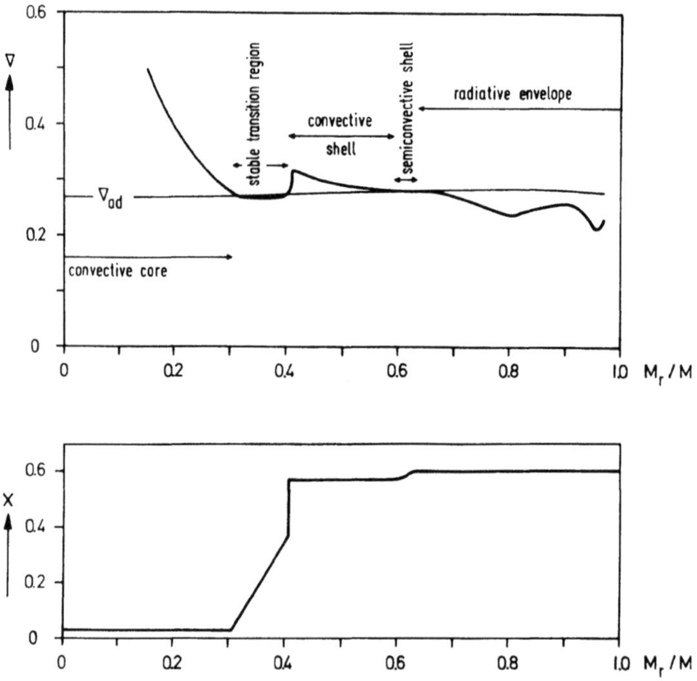

Fig. 9. Temperature gradient $\nabla = \mathrm{d}\ln T/\mathrm{d}\ln P$ and hydrogen content X as functions of depth (over a M_r/M-scale) of a 25 $M_\odot$ star after exhaustion of central hydrogen. ∇_{ad} is the value of ∇ in adiabatic regions. (The models were computed with the condition $\nabla = \nabla_{ad}$ for marginal stability in semiconvective regions).

A very careful discussion of the problem by Stothers (1969) came out just recently. He used the ratio of the numbers of blue supergiants to the number of red ones and found that this number is one or less for clusters with turnoff points at $M_v > -6$ but is 4 or bigger for clusters with brighter turnoff points. From this result Stothers (assuming neutrino losses) concludes, that for the first class the red supergiants are in the stage of central helium burning while for the younger clusters the blue supergiants and most of the red supergiants are in the stage of central helium burning but some of the red supergiants are in the stage of higher nuclear reactions like carbon burning. This is in agreement with our conclusion, that red supergiants can be in the stage of central helium burning. Stothers also concludes from statistical arguments that there must be central helium burning in the blue supergiant region. The question remains open, whether these blue helium burning stars are in a more advanced stage of helium burning and have moved along 'loops' toward the left or whether the transition region between helium core and hydrogen rich envelope is different for different stars of the same cluster – sharper for the red supergiants, broader for the blue ones.

Acknowledgements

I thank Dr. L. Lucy for valuable discussions and Mr. W. Sanders for improving the English of the manuscript.

References

Arp, H.: 1967, *Astrophys. J.* **91**, 149.
Giannone, P., Kohl, K., and Weigert, A.: 1968, *Z. Astrophys.* **68**, 107.
Hayashi, C., Cameron, R. C.: 1962, *Astron. J.* **67**, 577.
Hayashi, C. and Cameron, R. C.: 1962, *Astrophys. J.* **136**, 166.
Hodge, P. W.: 1966, *Galaxies and Cosmology*, McGraw Hill, New York.
Hofmeister, E.: 1967, *Z. Astrophys.* **65**, 164.
Iben, I.: 1966, *Astrophys. J.* **143**, 516.
Kippenhahn, R.: 1969, *Astron. Astrophys.* **3**, 83.
Kippenhahn, R. and Weigert, A.: 1967, *Z. Astrophys.* **65**, 251.
Kotok, E. V.: 1965, *Soviet Astron.–AJ* **9**, 948.
Kotok, E. V.: 1966, *Soviet Astron.–AJ* **10**, 254.
Meyer-Hofmeister, E.: 1969, *Astron. Astrophys.* **2**, 143.
Paczynski, B.: 1967, *Acta Astron.* **17**, 355.
Stothers, R.: 1963, *Astrophys. J.* **138**, 1074.
Stothers, R.: 1969, *Astrophys. J.* **155**, 935.
Stothers, R. and Chin, C.-W.: 1969, private communication.
Wilde, K.: 1969, thesis Göttingen.
Wildey, R. L.: 1966, *Colloquium on Late Type Stars* (ed. by M. Hack), Trieste, p. 405.

NUCLEOSYNTHESIS IN THE MAGELLANIC CLOUDS AND THE GALAXY

GEOFFREY BURBIDGE
Dept. of Physics, University of California, San Diego, Calif., U.S.A.

It has been clear for many years that the bulk of the chemical elements making up the solar system and the stars in our Galaxy were not formed in a single event of nucleosynthesis but were generated in a variety of different processes (Burbidge, Burbidge, Fowler, and Hoyle, 1957 hereinafter referred to as B^2FH). Only the helium could have been synthesized in a big bang, if such a cosmological event occurred, and the present evidence suggests that the bulk of the Galactic helium was not synthesized in this way (Burbidge, 1969). Thus, it appears that the elements were synthesized in a variety of normal stars at different stages in their evolution, and also probably in more massive stars which may have been present at some epoch. If the composition of our Galaxy is explained in this way, it might be reasonable to assume that a particular set of processes occurred which are unique to our Galaxy, and that therefore its element abundance curve is different from that of all other galaxies. A test of this hypothesis would be to investigate the chemical compositions of other galaxies which have not gone through the same evolutionary processes. The nearest galaxies, and thus those which are easiest to study, are the Magellanic Clouds.

However, this type of investigation will not achieve the required result unless the external galaxies studied are really separate systems which have had a completely different evolutionary history. What is plausible for the Magellanic Clouds?

They are dwarf irregular systems situated at distances of order 60 kpc from the Galaxy, and they are separated from each other by a distance which is less than, but comparable with their distance from the Galaxy. The radial velocities of the Clouds with respect to the Galactic center are very small. The Magellanic Clouds are very similar to the very large population of dwarf galaxies which populate the Universe. They may be (a) satellites which are bound to our Galaxy, and were formed with it, (b) field galaxies which are simply in the vicinity of the Galaxy, or (c) galaxies which were formed with our Galaxy but which have positive energy and are being ejected from it.

Holmberg has suggested that there is an excess of dwarf galaxies in the vicinity of bright galaxies and this might suggest that satellite galaxies tend to be formed preferentially in association with larger galaxies.

However, in the case of the Galaxy and the Clouds it appears to me that we have no real evidence as to whether (a), (b), or (c) is more likely. If (a) or (c) is correct, early processes of nucleosynthesis in our Galaxy and the Magellanic Clouds may well have been common to both systems. On the other hand, if (b) is correct, studies of the composition of the Clouds may help us to understand the processes of nucleosynthesis on the larger scale.

Muller (ed.), The Magellanic Clouds, 156–162.

We review next the chemical composition of the Galaxy and the Magellanic Clouds. Since the original discussion of B^2FH, more evidence has accumulated that elements are currently being synthesized in various types of star in our Galaxy. Reviews of the data have been given successively by Burbidge (1962), by Cayrel and Cayrel de Strobel (1966), by Pagel (1968), and by Unsold (1969).

To establish that elements are being synthesized in stars it is necessary to show:

(a) That under appropriate conditions nuclear reactions can be shown to occur that will build from the elementary particles, baryons and fermions, by one process or another, all of the elements in the periodic table. By laboratory experiment, extrapolation of experimental results, and nuclear theory this has been established.

(b) That stars proceed through evolutionary sequences such that the temperature, density, and compositions in their interiors allow these processes to occur. Theoretical studies of evolution of stars and color-magnitude diagrams of star clusters have indicated that suitable conditions will naturally occur. As the later stages of evolution are reached, stellar evolution speeds up very rapidly and we cannot expect to observe many stars which are passing through these stages of evolution. Thus so far comparison between theory and observation of stellar evolution has only taken us to the horizontal branch of the Hertzsprung-Russell diagram. However, there is no doubt that in the later phases of stellar evolution and in explosive events, rapid element synthesis will occur.

(c) That stars in the Galaxy exist in which element synthesis is taking place currently, or has taken place in these stars since they formed. Many abundance analyses exist and many stars have been found in which this evidence is found. The situation was originally reviewed by B^2FH. Analyses carried out in the decade since this work was completed further strengthen the evidence. Stars in which hydrogen burning, helium burning, the *s*-process, and the production and/or destruction of lithium is going on are well known. We know that generation of the elements in the iron peak (the original *e*-process) and nucleosynthesis beyond it involving rapid neutron capture processes must go very rapidly in stars which rapidly proceed to a catastrophic phase. Thus we cannot expect to see stable stars which can be directly identified with these processes. It was originally expected that supernova remnants would show the effects of the *e*-process and the *r*-process. The situation here is still not clear. It was originally proposed that Cf^{254} was responsible for supernova light curves of type I and this isotope can only be produced in an *r*-process. Zwicky has shown that there is considerable variation in supernova light curves and this, from an observational standpoint, is an objection to the idea that a single transuranic isotope is responsible. However, detailed *r*-process calculations have shown that other isotopes will contribute to the light curve so that the case for *r*-process isotopes explaining supernova light curves is still very strong. A very different mechanism proposed by Morrison and Sartori (1966, 1969) is open to serious objections. However, the Crab nebula, the only well studied supernova remnant, does not show abundance anomalies among the lighter elements which would indicate that drastic nucleosynthesis has occurred, but attempts are being made to detect evidence for the *r*-process from soft γ-ray lines

(Clayton and Colgate, 1969). Since the original work of B^2FH there still has been no progress in identifying lines or bands found in the spectra of supernovae in external galaxies, though Greenstein has frequently stated, without supporting evidence, that he has some supernova spectra in which Fe is prominent. This, if correct, would be strong evidence in favor of the *e*-process occurring in supernovae.

Overall, therefore, the evidence is overwhelming that nucleosynthesis involving a large fraction of the elements is currently going on in stars, some of which we can study individually in our own Galaxy.

Little evidence from individual stars in the Magellanic Clouds is available simply because they are so much farther away that, in all but a few cases, it is impossible at present to do comparable abundance analyses. However, studies of the nuclei of planetary nebulae (Faulkner and Aller, 1965) suggest that these have compositions similar to those in our own Galaxy and may thus have gone through similar evolutionary processes. We shall come to other individual observations later.

The second aspect of the theory of element synthesis in stars is to account quantitatively for all of the isotopes in the solar system on the basis that they have been made in stars at different stages of their evolution, treating the solar system as a typical object with an age of about 5×10^9 yr. More generally we must ask whether the composition of our Galaxy, of the Magellanic Clouds, and perhaps of all galaxies is made up of the products of element synthesis which has occurred in known types of star.

The arguments that were originally made by B^2FH were concerned solely with our own Galaxy. We now reconsider the situation, both for our Galaxy and the Magellanic Clouds. The original idea was that the protogalaxy condensed out of a cloud of hydrogen, and by stellar nucleosynthesis the elements were built from hydrogen. This would imply that the first generation of stars which formed contained no elements other than hydrogen, and there was a progressive enrichment in heavier elements as a function of the age of stars and their dynamical characteristics. It was hoped therefore that it might be possible to find pure hydrogen stars which would be the oldest low-mass stars, and that there would be a reasonably good correlation of composition with age. However, even in the 1957 paper complications in this simple picture were considered.

What is the observational situation at present? No pure hydrogen stars have ever been found anywhere. If there are really none now in existence in our Galaxy, this either means that the initial condensations of pure hydrogen were all massive enough so that they have evolved in a time short compared with the age of our Galaxy or any other galaxy in which their absence can be established. For the Magellanic Clouds whether or not pure hydrogen stars are present is not known. As will be discussed later the Clouds, if treated as separate systems, are likely to be at an earlier evolutionary stage than our own Galaxy, so that on some hypotheses they may be more likely to be present.

There are stars in our own Galaxy which have very low metal abundances. These are the classical old extreme population II halo stars. The easiest parameter to measure using coarse techniques is the Fe/H ratio and one can use UBV photometry or qualitative studies of very low dispersion spectra. Such studies by Eggen *et al.* (1962) and Dixon (1966) have led to the conclusion that the halo population (stars similar to those in the globular clusters) with large eccentricities, low angular momenta and high velocities have low Fe/H ratios (the ratio relative to the sun being $\gtrsim 5$); on the other hand, the disk population with low eccentricities, high angular momenta, low inclination tend to have Fe/H ratios of the order or sometimes greater than that of the sun. From these results it has been concluded that after an initial burst of element formation in a time certainly less than 10^9 yr after the bulk of the stars in the Galaxy arrived on the main sequence, there was steady enrichment of heavy elements in the disk, but in the halo element synthesis effectively stopped after the globular clusters were formed. However, the studies of some old galactic clusters show that they are often as abundant in heavy elements as stars which are much younger. This points to a large amount of element synthesis taking place in the early history of the Galaxy, though the effect must have been patchy since there are disk stars with low metal abundances. There has been considerable discussion of the abundance differences between different heavy elements in disk stars as they point to different nuclear processes. However, these analyses only refer to our Galaxy and we have no information regarding these more complicated results for the Magellanic Clouds.

What abundance results are available for the Clouds? The only conventional abundance analysis I am aware of is that of the supergiant HD 33579 made by Przybylski (1965). He has attempted to measure the abundances relative to hydrogen of Fe, Cr and Ti. By comparing the results with those obtained for the galactic supergiant α Cygni he has found that the abundances agree to within a factor of about 2, and this means that the abundances are essentially the same. Other results which are available are based only on qualitative investigations of spectra. Such investigations of B 0 to K 5 supergiants have been carried out by Feast *et al.* (1960) and by Code and Houck (1958) for a very luminous star, and comparison of the spectra which have been taken at dispersions of 30–90 Å/mm show no obvious differences in relative intensities of the lines compared with population I galactic disk stars obtained at similar dispersions. Thus, the qualitative conclusion is that for the most abundant elements up to the Fe peak and in some cases as far as heavy elements such as Ba there is no appreciable difference in composition. We have already mentioned the gaseous nebulae. A number of investigations of the chemical composition of S Doradus and one or two other nebulae have been made, and while there are some small differences, the abundances of hydrogen, helium, and oxygen appear to be about the same as they are in similar nebulae in our own Galaxy.

It would therefore appear to be the case that the bulk of the matter which is easily studied in the Magellanic Clouds – and it must be remembered that we have only been able to look at the bright end of the stellar luminosity function in the Clouds, and this very crudely – contains elements with similar total and relative abundance to that in the disk of our Galaxy. There is no way at present of obtaining spectra of fainter stars which presumably make up the bulk of the mass of the Clouds. To study these we must attempt to interpret the color magnitude diagrams of some of the star clusters. Investigations of a number of clusters have been made. They have been called 'globular-like' to distinguish them from Galactic globular clusters. The situation has been investigated in detail and summarized by Gascoigne (1966). He has pointed out that there are at least two clusters, NGC 1466 and NGC 2257, which closely resemble galactic halo globular clusters in their color-magnitude diagrams and which must therefore, without further evidence, be supposed to be old and metal-poor. There are also a number of clusters in the Small Magellanic Cloud which seem to be young and which he therefore supposes have normal metal content. Then there is an appreciable fraction of clusters whose color-magnitude diagrams are different from any seen before, and some which may be globular clusters at an early stage in their evolution.

Thus, the conclusion from this work is that there may very well be a range of element abundances in the Magellanic Clouds similar to that in our Galaxy. Since the galaxies are of very different types with different masses, and since the fractions of gas to condensed matter are different, it is commonly supposed that their evolutionary histories have been quite different. If this is the case, why have the nucleosynthesis processes been so similar? The first point to make is, of course, that we are not really sure whether the scanty chemical composition studies of the Clouds warrant this interpretation. It might be, for example, that there is a comparatively large population of low mass stars which are metal poor. If we ignore this possibility we must speculate that the Magellanic Clouds have been through similar nucleosynthesis processes to those in our Galaxy, i.e. an early burst of element synthesis occurring over a short time scale, leaving a somewhat patchy distribution of abundances, but the building of the elements up to the Fe peak in a comparatively large fraction of the mass of the galaxy. If the indirect evidence from the globular-like clusters is to be believed, some clusters are very metal poor and thus have avoided contamination from the main element synthesis process.

One obvious explanation of this state of affairs is that the Magellanic Clouds were formed originally with our own Galaxy, and in the early phases of nucleosynthesis the matter was not separated. This requires that the Clouds fragmented from our Galaxy and were in effect ejected from it, i.e. the data are compatible with possible origins (a) and (c) mentioned in the beginning of this paper.

What little we know of the dynamics of the Galaxy and the Clouds does not rule this out, but such a proposal immediately brings to the fore the unsolved problem of the formation of the Galaxy. Thus we have no way at present of deciding whether this is a reasonable hypothesis or not. As was mentioned earlier, it is very important in this connection to determine the predominant composition of other galaxies, and especially those *outside* the Local Group, so that there is no possibility that we are looking at different objects which went through their initial nucleosynthesis together. However, at present very little data on more distant galaxies is available. All that one normally obtains is the integrated spectrum of the nucleus of a galaxy. From such spectra attempts are made to obtain an estimate of the stellar population and luminosity function of the stars in the central region, and also from the Doppler broadening of the integrated spectrum to obtain a measure of the velocity dispersion of the stars, and hence through the virial theorem an estimate of mass. In general it has been *assumed* that the compositions of the stars are the same as those in our own Galaxy, in order that the number of possible variable parameters can be reduced. Even then attempts to study the stellar population have been restricted to the nucleus of M 31, M 32 (both Local Group members), and a few of the brighter ellpiticals like NGC 3379 and some Virgo cluster members. The second method has been used also only for bright ellipticals. The fact that the integrated broadened spectra can be roughly explained as the effect of integrating the spectra of galactic stars of several spectral types after a broadening function is applied does not allow one to conclude with any certainty that the chemical compositions of these galaxies are the same as that in our Galaxy. Thus, the recent statement of Unsold (1969) that, from evidence of this kind, the conclusion should be drawn that the H/metal ratio and the relative abundances of the prominent elements are the same in most galaxies in the Hubble sequence to within rather narrow limits, is much too strong. In rebuttal it must first be stated that we only have evidence of this type from a very small number of galaxies outside the Local Group, and second, a qualitative perusal of the integrated spectra of many galaxies shows that the prominent stellar features, the H and K lines of Ca II and the G band, do not always appear to be similar. This may or may not mean that there are abundance differences. A different method of approach is to investigate the emission lines in the nuclei of galaxies. One of the most striking features found is that the ratio of $H\alpha$/[N II] 6583 varies as a function of position in many spiral galaxies. In the nuclear region the [N II] lines are much stronger relative to $H\alpha$ than in the outer parts (Burbidge, and Burbidge, 1962). It was originally thought that this might be an excitation effect, and less probably due to abundance differences. Recently Peimbert (1968) has re-investigated this question and concluded that it is most likely an abundance effect. While this result is not certain, if correct, it would mean that nitrogen has been synthesized more rapidly and to a greater extent in the nuclear regions of spiral galaxies than in the outer parts. Thus it would indicate that for nitrogen at least enrichment in a galaxy is a function of position, and that the abundance of a light element is different in different spiral galaxies. There are also galaxies with highly peculiar spectra which suggest that abundance differences may be present. A good example is the compact system investigated by Sargent (1968) in which very strong emission lines of iron are seen. This may be an excitation effect, but it may also indicate that this extragalactic system is abnormally rich in Fe.

Thus, on the basis of the rather scanty evidence, one may tentatively conclude that the compositions of galaxies in general are not all likely to be the same, even as far as the Fe/H ratilo and the common elements are concerned.

To summarize, there is not enough evidence on the chemical composition of the Magellanic Clouds when compared with our Galaxy for a detailed theory of the chemical evolution of the Clouds to be developed at present. The evidence as far as it goes is compatible with the view that much of the material of the Clouds went through a considerable amount of nucleosynthesis early in its history. The Clouds could once have been part of the Galaxy, or they could have formed as satellites when the proto-galaxy condensed. However, we do not know at present whether the Galaxy formed from a diffuse state or evolved from a highly condensed object. If the Clouds have a very different evolutionary history from the Galaxy, and if they are simply to be treated as representative of galaxies in the field, then their rather similar composition requires us to argue that the processes of nucleosynthesis early in the development of galaxies can be rather similar. However, before any firm conclusions can be drawn a much wider sample of galaxy compositions must be obtained, and preliminary evidence suggests that, in general, there are composition differences between galaxies. Nucleosynthesis processes which give rise to a considerable degree of enrichment of the heavier elements in the early history of a galaxy require the rapid evolution of massive or supermassive stars. Such processes must have occurred in our Galaxy and probably also in the Clouds. Theoretical investigations of nucleosynthesis in such objects are under way (Wagoner, 1969; Arnett, 1969).

The general problem of the chemical evolution of a galaxy is tied closely to the problem of galaxy formation which remains unsolved.

Acknowledgement

Theoretical research in astrophysics is supported at UCSD by the National Science Foundation, and by NASA under grant NGR-05-005-004.

References

Arnett, D.: 1969, preprint.
Burbidge, G. R.: 1962, *Ann. Rev. Nuclear Sci.* **12**, 507.
Burbidge, G. R.: 1969, *Comments on Astrophysics Space Sci.* **1**, 101.
Burbidge, E. M. and Burbidge, G. R.: 1962, *Astrophys. J.* **135**, 694.
Burbidge, E. M., Burbidge, G. R., Fowler, W. A., and Hoyle, F.: 1957, *Rev. Mod. Phys.* **29**, 547.
Cayrel, R. and Cayrel, G. de Strobel: 1966, *Ann. Rev. Astron. Astrophys.* **4**, 1.
Clayton, D. and Colgate, S. 1969, preprint.
Code, A. D. and Houck, T.: 1958, *Publ. Astron. Soc. Pacific* **70**, 261.
Dixon, M. E.: 1966, *Monthly Notices Roy. Astron. Soc.* **131**, 325.
Egge, O. J., Lynden-Bell, D., and Sandage, A. R.: 1962, *Astrophys. J.* **136**, 748.
Faulkner, D. J. and Aller, L. H.: 1965, *Monthly Notices Roy. Astron. Soc.* **130**, 393.
Feast, M. W., Thackeray, A. D., and Wesselink, A. J.: 1960, *Monthly Notices Roy. Astron. Soc.* **121**, 337.
Gascoigne, S. C. B.: 1966, *Monthly Notices Roy. Astron. Soc.* **134**, 59.
Morrison, P. and Sartori, L.: *Phys. Rev. Letters* **16**, 414.
Morrison, P. and Sartori, L.: 1969, preprint.
Pagel, B. E. J.: 1968, *Q. J. Roy. Astron. Soc.* **9**, 401.
Peimbert, M.: 1968, *Astrophys. J.* **154**, 33.
Przybylski, A.: 1965, *Nature* **205**, 163.
Sargent, W. L. W.: 1968, *Astrophys. J.* **152**, L31.
Unsöld, A.: 1969, *Science* **163**, 1015.
Wagoner, R. V.: 1969, preprint.

IMPORTANCE OF MAGELLANIC CLOUDS STUDIES FOR EXTRA-GALACTIC WORK

E. MARGARET BURBIDGE

Dept. of Physics, University of California, San Diego, Calif., U.S.A.

Abstract. The apparently early evolutionary state of the Magellanic Clouds, while they are coeval with the Galaxy in their earliest star condensations, may be due to perturbation by the Galaxy as well as to their low mass and low density. Whereas classical barred spirals are probably characterised by large angular momentum, the Clouds may owe their 'barred' and asymmetrical distribution, including departure of the centre of rotation from the geometrical centre, to local fluctuations in H I density and rate of star formation at different epochs. Comparison is made with some other asymmetrical galaxies.

1. Introduction

The conventional scheme of galactic evolution envisages a protogalaxy forming and having an initial stage of considerable asymmetry, a high content of uncondensed matter, and turbulent or disordered motions. According to its initial conditions of mass, density, angular momentum, turbulence, and magnetic field, one envisages such a protogalaxy progressing to a state in which all or most of the material has condensed into stars, rotational symmetry has been achieved, and non-circular motions have died out except for those frozen into the stellar population in the form of stellar orbits of high inclination and high eccentricity. Prendergast and Miller (1969) devised a computational scheme for following the evolution of a galaxy represented by 10^5 mass points, with star formation going as some function of the gas density (e.g. as the square of the density). Some of the early forms produced by these computations during the evolution of a simulated galaxy of this kind look remarkably like some of the well-known irregular galaxies such as NGC 4038-9 (cf. Burbidge and Burbidge, 1966).

According to these considerations, the Magellanic Clouds should be at a relatively early stage in their individual evolutionary histories, since they contain much uncondensed material and young stars, and are unsymmetrical in form (see discussion by Arp (1964)). Yet the oldest star condensations in them seem to be coeval with the oldest stars in the Galaxy, and with those in the intergalactic globular clusters around the Galaxy and in the pure population II dwarf systems in the Local Group. The nearness of the Magellanic Clouds, which makes it possible to study individual stars in coeval groups and star clusters in some detail and to date them in years, thus makes the Clouds of great importance in our attempts to understand the evolution of galaxies. The evidence points to there having been local bursts of rapid star formation interspersed with quiescent periods, presumably as a result of local variations in H I density.

Since the Clouds contain very old stars, one might argue by analogy and suggest

Muller (ed.), The Magellanic Clouds, 163–168. *All Rights Reserved.*

that all highly irregular galaxies, no matter how early they appear to be in their own evolutionary histories, may likewise contain a substratum of very old stars. In the more distant galaxies there is no way yet to check this.

The nearness of the Clouds, however, has its drawbacks, also; it is not easy to delineate their boundaries and separate them from the Galaxy, and it is also hard to study their dynamical relationship with the Galaxy.

In the succeeding sections I shall briefly discuss the relationship of the Clouds with each other and with the Galaxy, their asymmetry and rotation in relation with other asymmetrical systems, their relationship with barred spirals, and finally some further general considerations on galactic evolution.

2. Relationship with the Galaxy

The small line-of-sight velocity of the Clouds relative to the center of the Galaxy makes it probable that they are bound satellites of the Galaxy. The actual value of this velocity depends on the true value of the sun's circular velocity but is not much greater than 60 km/sec for the LMC and less for the SMC. The Clouds have a common envelope of hydrogen, and therefore mixing that can occur in the gaseous content of both can also occur to some extent between the two, although the time scale may be long. Whether there is a connecting bridge or arm to the Galaxy is more uncertain, but it appears likely that there can have been an interchange of gas between all three, including the products of nucleosynthesis in stars and in central explosive events in our Galaxy, as discussed by G. Burbidge in this volume.

In discussing the dynamics of the Large Magellanic Cloud, a point that merits more attention is that raised by Thackeray (1964) concerning the effect on its rotation curve produced by the gravitational field of the Galaxy.

Little quantitative work has been done on rotation in close binary or multiple systems; the perturbing effect of M 32 on the rotation of M 31 was considered by Schwarzschild (1954) in an early attempt to estimate the mass of M 32, and the effects of the Magellanic Clouds on our Galaxy was discussed by Avner and King (1967) who found that these could cause a bending of the outer edges of the galactic plane similar to that observed. If the mass of the LMC is $3 \times 10^9\, M_{\odot}$, then at a distance of 55 kpc from the center of our Galaxy the gravitational force due to the Galaxy is of the same order of magnitude as that exerted by the LMC 8 kpc from its center (the diameter of the LMC is about 16 kpc according to de Vaucouleurs (1955)).

The shape of the Clouds may be considerably affected by this gravitational interaction. The three-dimensional form of the Clouds is not known and it has often been assumed that the LMC is a highly flattened object seen nearly face-on, so that the measured line-of-sight velocities are multiplied by a large factor to give the presumed rotational velocities. An argument that has been adduced in favour of this flattening is the smallness of the velocity dispersion, which has been taken to mean that there is very little turbulent motion to support material out of the equatorial plane. But in view of the irregular form of the LMC and the strong gravitational perturbation by the Galaxy, I believe the orientation of the plane of the LMC is quite uncertain, and the factor that has been used to correct the observed rotation curve is probably much too large. The mass will then be considerably less than $10^{10}\,M_{\odot}$.

3. Asymmetry and Rotation

I have already referred to asymmetry as a property which suggests an early evolutionary stage. In a few rotation periods a system possessing angular momentum will begin to progress toward axial symmetry; a large angular momentum per unit mass may favour the formation of barred spirals. Burbidge *et al.* (1963) studied and published photographs of a number of highly irregular galaxies which all had much uncondensed gas and a young stellar population with hot stars to ionize the gas. These galaxies possessed velocity gradients which could be interpreted as rotations, and their irregular and generally elongated forms of large dimension would be expected eventually to wind up into spiral or barred spiral forms. An important distinction between these objects and the Magellanic Clouds lies in their very different dimensions, the Clouds being very much smaller.

A comparison between the Magellanic Clouds and various asymmetrical irregular-spiral galaxies was made by de Vaucouleurs (1954) as part of the argument for the Clouds being barred spirals. Figure 1 shows three more asymmetrical but otherwise not highly irregular galaxies, which have a much greater resemblance to ordinary spirals than to barred spirals, showing that asymmetry can be found in these cases also. NGC 4234 is a well-known galaxy; VV112 was catalogued by Vorontsov-Velyaminov (1959) but has not been studied spectroscopically.

MCG 4-31-14, in Vorontsov-Velyaminov's *Morphological Catalogue of Galaxies*, is a particularly interesting case. The arms on one side look like ordinary spiral arms in a galaxy whose other half, the opposite side of the "nucleus", is missing, almost as though it were obscured by a dense dust cloud. A spectrogram of this object was obtained in the photographic wavelength region with the prime-focus spectrograph on the Lick 120-inch telescope, and the velocities were measured along the major axis of the extended [O II] λ 3727 line. Strangely, the velocity gradient is steepest at the end away from the 'nucleus'; around the latter the rotation curve is fairly flat with a mean value of +2560 km/sec. From this, with a Hubble constant of 75 km/sec per Mpc, the distance of the object is about 34 Mpc and the diameter is 12.9 kpc. The total range of velocity, which needs to be checked by measurements at Hα with greater dispersion than this preliminary observation, is about 250 km/sec. The apparent nucleus may thus not be the region of highest density; it may appear bright only because it is the region where a large burst of star formation has recently occurred.

Finally, some emphasis has been placed on the rotation curve in the Large Magellanic Cloud not being symmetrical about the apparent optical center. This phenomenon has been found in a number of small galaxies of fairly low mass, for example, NGC 3310, NGC 4490, and NGC 4605 (Bertola, 1966; Walker and Chincarini, 1967). Here, as in MCG 4-31-14, it is perfectly feasible that the region of greatest total mass

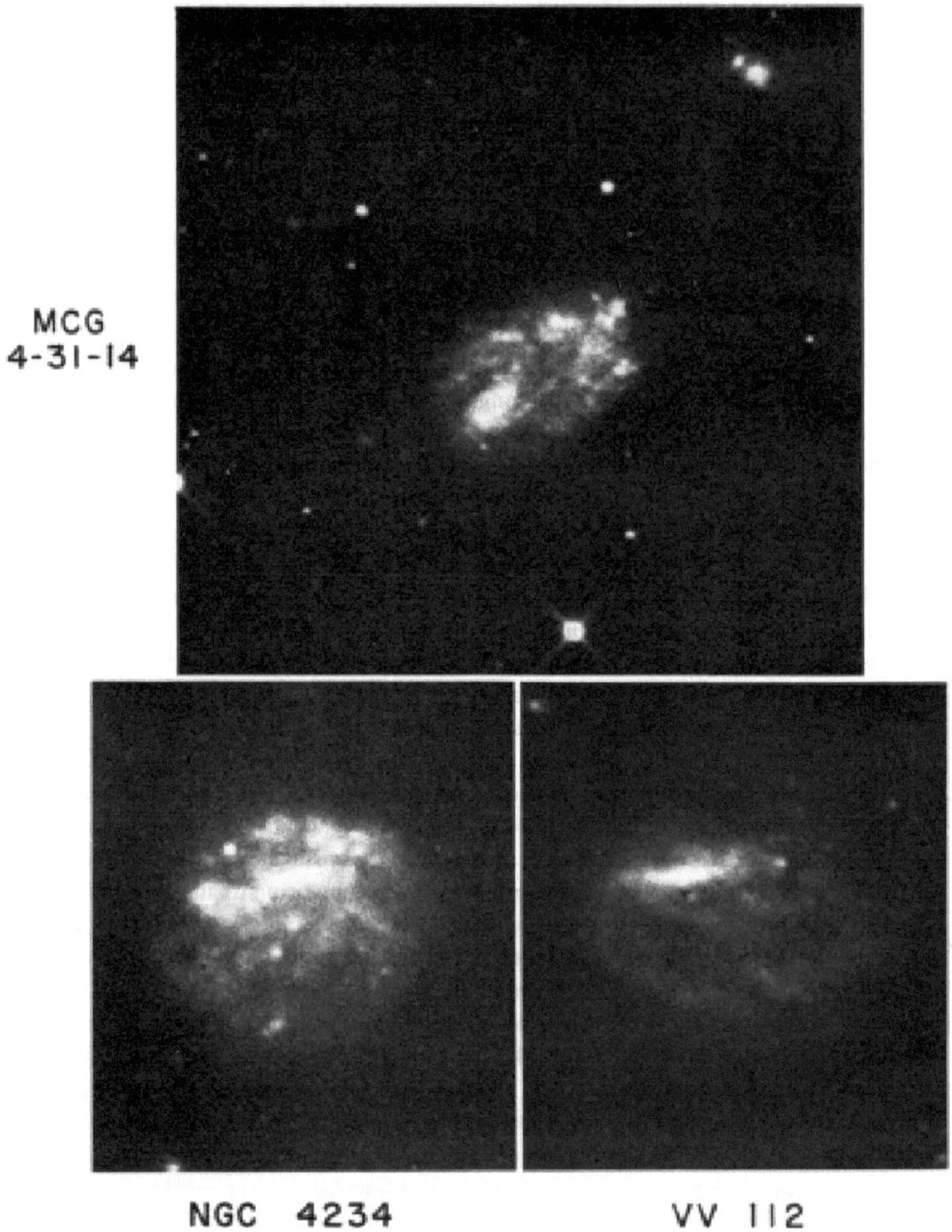

Fig. 1. Three asymmetrical spiral galaxies (Lick 120-inch and McDonald 82-inch photographs; 103a0 emulsion without filter).

density may not coincide with the centre of greatest activity in star formation in the recent past.

4. Barred Spirals and Angular Momentum

There has been considerable discussion of the Clouds in relation to barred spirals. It has been thought likely that the barred spirals have greater angular momentum per unit mass than normal spirals, without much supporting evidence. The only full rotation curve that has been determined for a 'classical' barred spiral is the case of NGC 7479 (Burbidge *et al.*, 1960), and it is true that here the curve is linear over the large dimension of the bar (18 kpc) and the angular momentum is large. Also, rapid rotation of the nuclear region is a characteristic feature found in several barred spirals.

In the LMC it is very dificult to form any meaningful estimate of the angular momentum. The orientation of the plane is uncertain and the mass distribution is poorly determined. Uncertainty in the mass distribution in the outer parts of even regular Sc galaxies with well-observed rotation curves makes the determination of angular momentum highly uncertain. The outer parts, where the galaxy is faint and spectra hard to record, contribute heavily to the total angular momentum despite their low density, and fitting an empirical rotation curve following a chosen force law such as the Bottlinger-Lohman or Brandt-Belton formulae implies assumptions about the mass distribution.

The Magellanic Clouds do not resemble the classical barred spirals but rather those dwarf unsymmetrical objects in which elongated structures appear. Their form may thus be due to something quite different from possession of a large angular momentum.

5. Conclusion: Galactic Evolution

In the previous sections reasons have been given for thinking that the Magellanic Clouds are not far advanced along an evolutionary path toward a regular, axially-symmetric form, despite the fact that they are as old as our Galaxy, judging by their oldest stars. Holmberg (1964) has shown that the evolutionary stage of a galaxy (irrespective of its age in years) is correlated primarily with density, star formation having gone fastest in the high-density ellipticals, both giants like M 87 and dwarfs like M 32. Despite the uncertainty in the masses of the Clouds, they are clearly of low density. They do not have high-density nuclear regions, and consequently it is unlikely that a violent nuclear explosion could ever have occurred in either of them.

The perturbing effect of our Galaxy on the dynamics of the Clouds may hinder their approach to a symmetrical, regular form, in addition to the tendency in any case for low-mass galaxies to have a low rate of galactic evolution. During the approach to a steady state of symmetrical mass distribution, the random shifting of areas of temporarily high H I concentration may lead to the occurrence of pockets and areas of most rapid star formation well away from the geometrical center. The bar-like structure may be related to an asymmetrical distribution of past areas of greatest star formation rather than to a similarity to the classical large barred spirals which probably all have large angular momentum.

References

Arp, H. C.: 1964, in F. J. Kerr and A. W. Rodgers (eds.), 'The Galaxy and the Magellanic Clouds', *IAU-URSI Symp.* **20**, 219.
Avner, E. S. and King, I. R.: 1967, *Astron. J.* **72**, 650.
Bertola, F.: 1966, *Contr. Obs. Astrophys. Asiago*, Nos. 172, 186.
Burbidge, E. M. and Burbidge, G. R.: 1966, *Astrophys. J.* **145**, 661.
Burbidge, E. M., Burbidge, G. R., and Hoyle, F.: 1963, *Astrophys. J.* **138**, 873.
Burbidge, E. M., Burbidge, G. R., and Prendergast, K. H.: 1960, *Astrophys. J.* **132**, 654.
Holmberg, E.: 1964, *Arkiv Astron.* **3**, 387.
Prendergast, K. H. and Miller, R.: 1969, private communication.
Schwarzschild, M.: 1954, *Astron. J.* **59**, 272.
Thackeray, A. D.: 1964, in F. J. Kerr and A. W. Rodgers (eds.), 'The Galaxy and the Magellanic Clouds', *IAU-URSI Symp.* **20**, 380.
Vaucouleurs, G. de: 1954, *Observatory* **74**, 23.
Vaucouleurs, G. de: 1955, *Astron. J.* **60**, 126.
Vorontsov-Velyaminov, B. A.: 1959, *Atlas of Interacting Galaxies*, Moscow.
Walker, M. F. and Chincarini, G.: 1967, *Astrophys. J.* **147**, 416.

PART IV

PROSPECTS FOR FUTURE RESEARCH

POSSIBILITIES OF NARROW-BAND PHOTOMETRY, ESPECIALLY FOR MAIN-SEQUENCE PROBLEMS

BENGT STRÖMGREN
Astronomisk Observatorium, Copenhagen, Denmark

I would like to discuss briefly the possibilities of narrow-band photometry. The discussion will be based on the experience that has been gained, through the last several years, in research in our own galaxy with these methods. I would like to talk especially about main-sequence problems.

There are exciting possibilities for application of narrow-band photometry also to subgiants, giants, and horizontal branch stars, but the possibilities have not been as fully explored for these types of stars through research in our own galaxy as have the possibilities for main-sequence stars, and I shall therefore limit the discussion to the main-sequence stars.

In research on the Magellanic Clouds we have already heard of impressive results that have been obtained for supergiants and we are going to hear today about work on cepheids. In both cases conclusions can be drawn concerning the basic parameters of these stars, masses, ages and chemical compositions. It is clear, however, from research on stars in our own system that it is important to pursue the studies for main-sequence stars. We have here a phase where the chemical composition is more likely to be close to the original composition, and we are concerned with a phase where the theoretical calculations can be carried out with higher accuracy. With regard to the cepheids particularly, there are special possibilities that will be discussed later today.

In regard to the problem of determining ages and chemical compositions we know from the applications of narrow-band photometry to objects in our own system that significant results can be obtained particularly for B, A and F stars, and I would like to discuss now what the possibilities might be to apply these same methods in the Magellanic Clouds. With a distance modulus close to 19 the observational problems certainly differ from those encountered in research in our own system and it is therefore of interest to see what one might expect from efforts to push to the limit with the large instruments that are expected to go into operation here in Chile in the next few years.

In discussing these methods, I would like to define narrow-band photometry as a method which uses at least one index with a band width that is less than a hundred Ångström. Here I shall discuss the method that is based on the combination of 4-colour uvby photometry and Hβ photometry (cf. e.g. B. Strömgren (1966)) and look at the possibilities of photoelectric photometry of this type for a distance modulus close to 19. We can estimate fairly well, assuming optimum values of quantum efficiency and transmission of the optics, what the number of photo-

Muller (ed.), The Magellanic Clouds, 171–173.

electrons would be in this type of photometry. If we look, first, at intermediate band photometry with a band width of 200 Å, we find for magnitude 20, with an aperture of 3.6 m, that the photocurrent amounts to about 20 photoelectrons per second. With Hβ photometry, i.e. with a bandwidth of the narrow band of 30 Å there is a loss of about 2 magnitudes compared to uvby photometry. In this connection I would like to emphasize that the difference between the limiting magnitudes in these two cases is not quite what is indicated by the ratio of the band width, because in the determination of colour index differences the individual intensities have to be measured to higher accuracy for equal accuracy in the indices. Therefore Hβ photometry is only one magnitude behind uvby photometry with regard to limiting magnitude.

Experience shows that one can comfortably measure intensities with photoelectric accuracy of 1% and better for a photocurrent of 20 photoelectrons per second. The integration times are long but it is quite feasible to integrate for the 10 or 15 min necessary to achieve the accuracy. In cases of special interest, with a 3.6 m telescope one would even press beyond this limit, but then the integrating times become so long that the number of stars that can be observed would be quite limited. In this connection, it is clear that the photographic image tube techniques will be of great importance.

Given the limitations just referred to, there are at least three applications of narrow-band photometry that appear feasible and promising.

First, there is the determination of ages of main-sequence stars of the spectral range B 2–B 9. Through photoelectric uvby and Hβ photometry the location of the star in the $[u-b]$-β-diagram is determined, and if the star is situated in the upper half of the main-sequence band, the age can be evaluated to an accuracy of 10–15%. For A 0-1 stars of the early group the situation is the same (see B. Strömgren (1966)). The stars in question are brighter than $M_v = 1^m$, and their ages range up to 300 million years. Systematic studies based on age determinations for this type of main-sequence stars should lead to results of interest with regard to the migration of stars in the Magellanic Clouds, and they might extend results such as those already obtained by C. Gaposchkin and S. Gaposchkin pertaining to cepheids.

Secondly, the study of the location of the ZAMS in the $[u-b]$-β-diagram for B 2-B 9 stars belonging to associations in the Magellanic Clouds should give valuable indications regarding the chemical composition of the stars in question. The location of the ZAMS is sufficiently sensitive to stellar heavy-element content Z that a determination of Z within about 15–20% appears possible. As in the corresponding galactic problem we are, however, left with the difficulty of separating, in the analysis, differences in the helium content Y and the heavy-element content Z.

Thirdly, if it should turn out to be possible to press accurate uvby photometry to $V = 21^m$ or beyond, a promising field of determining metal contents of individual main sequence stars (early F) from the index m_1 would open up. The observational difficulties are great, but the information gained regarding distribution of stellar chemical composition within a galaxy should be of considerable value.

DISCUSSION

Dr. Graham: I would like to mention briefly two important practical difficulties which one faces when one tries to observe stars as faint as this in the Magellanic Clouds. Firstly, the intense crowding of the field makes the isolation of individual stars difficult to achieve. Secondly, at the same time that the star is measured, the very irregular background must also be measured. The signal from this background will generally exceed that from the star by many times.

Dr. Strömgren: Dr. Graham has referred to a very important point. I would like to make three comments in reply to his remarks. The first concerns the question of the diameter of the diaphragm used in photoelectric multi-colour photometry. In work of the type contemplated it would be necessary to reduce the diaphragm-diameter by a considerable factor below what is now generally customary. Experience in photoelectric photometry where intensities in several bands are measured simultaneously suggests that this is indeed possible, when the reflector is technically first-class, and when the seeing is as good as is experienced on fine nights at La Silla. Such a reduction would very much ease both the sky-background and the star-blending problems. Secondly, we have undoubtedly in this connection to think seriously in terms of photographic photometry, in particular of the image-tube techniques discussed in this Symposium by Dr. Walker. With automatic, or semi-automatic, evaluation techniques the necessity of using averages of results from many plates in order to obtain sufficient accuracy should present no great problems, and the standard photographic and particularly the image-tube technique would then yield such information as to reduce the sky-background and star-blending problems. The latter technique should also be very useful in selecting the stars that lend themselves to photoelectric photometry. Finally I would suggest that photometry of the type that I have discussed might start with the less difficult areas of relatively low surface density such as occur in the wing of the Small Magellanic Cloud, and then be continued in areas of increasing difficulty.

IMAGE TUBE WORK ON THE MAGELLANIC CLOUDS

M. F. WALKER

*Observatorio Interamericano de Cerro Tololo**

It is most appropriate, in view of what I am going to discuss today, that we have in the audience Professor A. Lallemand, who is the father of the technique that I am going to report on. He has been working since 1935 on the development of electronographic image tubes, and the application of these devices in astronomical research has been producing important scientific results for about the last ten years. The full impact of this new technique has not yet been felt, but I am sure that in another twenty years everyone will appreciate what a truly revolutionary development in observational astrononomy Professor Lallemand has brought about.

The particular use of electronography that is of interest here is its application to stellar photometry. An extremely important characteristic of electronographic image tubes is that a linear relationship exists, over a large range in density, between the intensity of the light falling on the photocathode and the photographic density of the corresponding electronic image recorded in a special nuclear track emulsion. A few years ago Lallemand and his collaborators were able to show, using one of the plates that we took at Lick in 1959, that as a consequence of this linear response, one can obtain a linear relationship between the astrophotometer measurement and the intensity of the light of a star over a range of $2\frac{1}{2}$ magnitudes, starting at the plate limit. Since then, Dr. Kron and I have shown that one can obtain a linear relationship between star intensity and image density over a range of more than five magnitudes if one uses the proper readout technique to extract the information from the electronographic plates. Thus, the advantages of electronography for stellar photometry are:

(1) A photoelectric sequence is not needed. One requires only photoelectric magnitudes and colors of a few bright stars in the field to give the zero-point and color-equation, after which the magnitudes and colors of all other stars in the field may be measured directly on the electrograph down to the plate limit.

(2) Measurements may be made in much more crowded fields than is possible by ordinary photographic or photoelectric photometry, as well as in fields having variable background, due to the fact that we have a straight-forward manner of handling the necessary corrections as a result of the linear response of the tube.

(3) A telescope of moderate aperture may be used to reach the same detection limit as is attained photographically with the very largest reflectors, since the information gain of the electronographic image tube is, depending on the type of tube, 10 to 15 times the fastest photographic emulsion (Kodak baked IIa-0 at 4000 Å).

Three types of electronographic image-tube are presently available:

* Operated by the Association of Universities for Research in Astronomy, Inc., under contract with the National Science Foundation.

Muller (ed.), The Magellanic Clouds, 174–180.

(1) The electronic camera developed by Prof. Lallemand at the Paris Observatory.

(2) The modified, valve-type, electronic camera developed by Dr. Kron at the Lick and U.S. Naval Observatories.

(3) The 'Spectracon' developed by Professor McGee at Imperial College, London.

In the first two, the nuclear track emulsion is exposed to the photoelectrons in the same vacuum chamber as the photocathode. The Spectracon, on the other hand, is a permanently sealed tube having a very thin mica ('Lenard') exit window, through which about 75% of the photoelectrons pass to be recorded in the nuclear emulsion, pressed against it on the outside. Thus, while the gain of the Spectracon over photography is less than that of the 'classical' electronographic tubes, it has the advantage that it requires no laboratory work to prepare it for use at the telescope.

In coming to Chile, I planned to use the Spectracon for two programs:

(1) Spectroscopic observations. This program has not been carried out due to delays in the fabrication of the special spectrograph camera optics required.

(2) Direct electronography, with the Spectracon mounted at the focus of the 60-inch Tololo reflector. Two programs have been undertaken:

(A) Interference filter electronography of planetary nebulae.

(B) UBV filter electronography of globular clusters in the Magellanic Clouds.

It is this latter program that I shall discuss today.

The Spectracon, as its name implies, was designed for spectroscopic observations. As I have said, it has a very thin mica exit window, whose thickness is only about 4 μ. In order to obtain sufficient strength to enable this window to withstand atmospheric pressure, it is made in the form of a long, narrow slot, having dimensions of about 4.5×30 mm, of which about 4.5×23 mm may be used, due to the curvature of the window at the ends of the slot. While this shape is well suited to the recording of spectra, it amounts to a field of only about $1' \times 5'$ minutes of arc at the $f/7.5$ focus of the 60-inch reflector. Such a field is extremely small for stellar photometric studies, but it is just adequate for certain selected problems where one is dealing with very compact objects, such as the Magellanic Cloud clusters.

In the program, observations were restricted mainly to clusters for which photometric observations of the brighter stars already exist to provide the zero-point and color-equation of the electronographic magnitudes, Thus, the clusters observed were mostly those that have been studied earlier by Gascoigne, Arp, and others. The types of clusters included in the program were:

(1) Older clusters where the existing photometry does not go faint enough for us to tell whether the cluster is a population I or II object, that is whether it has a high or low metal content.

(2) Unusual objects like Hodge 11 that are alleged to have no counter-parts in the galaxy.

(3) Early-type clusters for a check on the distance moduli of the Clouds.

In addition, short exposures were taken of some of Arp's clusters to check the photoelectric sequences. The clusters observed in the program are listed in Table I. Some of these objects will be worked up in collaboration with Dr. Gascoigne, who is

TABLE I

Cloud	Cluster	Cloud	Cluster
SMC	Kron 3	LMC	NGC 1783
	NGC 121		NGC 1866
	NGC 330		NGC 2004
	NGC 361		NGC 2164
	NGC 419		NGC 2209
	NGC 458		Hodge 11
			NGC 2231
			NGC 2257

interested in particular problems involving certain of them, and who has promised to supply additional zero-point and color-equation standards for some of the others.

As I have said, the observations were made using broad-band UBV filters, and the electronic images were recorded on Ilford L4 nuclear track emulsion. The L4 is the finest grain nuclear track emulsion made by Ilford and, according to tests made by Dr. Kron, gives a photographic density proportional to the number of incident photoelectrons up to a photographic density of six. The density for a given exposure is lower than for some of the other nuclear track emulsions, but in fact one can detect stars down to the same magnitude limit in the same exposure time with this emulsion as with the ones giving a higher density. This, plus the fact that the response is linear, indicates that there is no loss in quantum efficiency with the L4 emulsion. The difference is simply that the information is displayed in a different way, with very fine grain and lower contrast. And because of the very fine grain of this emulsion, the number of bits of information that can be recorded is larger than for the other nuclear emulsions, which means that a longer exposure or integration time may be used so that fainter stars can be detected than with the others. With the coarser-grained Ilford G5 (whose granularity is still somewhat better than that of a Kodak IIa-0 plate), it is possible to expose with the Spectracon on the 60-inch for about one hour before the sky background becomes uncomfortably dense. However, with L4, it is possible to expose for at least three hours, and perhaps as long as six or eight hours. The magnitude limit on a three hour exposure depends, of course, on the quality of the seeing; for images with a half-peak intensity width of about 2″, it is about $V=23.0$ and $B=23.5$. When we recall that the faintest star observed photoelectrically by Dr. Baum with the 200-inch reflector was $V=22.9$ and $B=23.8$, it is clear that by means of electronography it is possible, using the present 60-inch telescope, to work on problems that otherwise would have to wait for the completion of the 150-inch reflectors.

Unfortunately, at present the results of this program are only qualitative as it has not been possible as yet to commence the quantitative measurements of the electrographs. Nevertheless, certain conclusions are possible from a simple inspection of the electrographs.

Figure 1 compares the photographic and electronographic observations of the cluster Kron 3 in the SMC. The upper picture in the figure is Gascoigne's limiting-exposure photograph of Kron 3 in yellow light, using the 74-inch Stromlo reflector. The exposure was 60 min on a Kodak 103a-D plate and through a Chance OY4 filter.

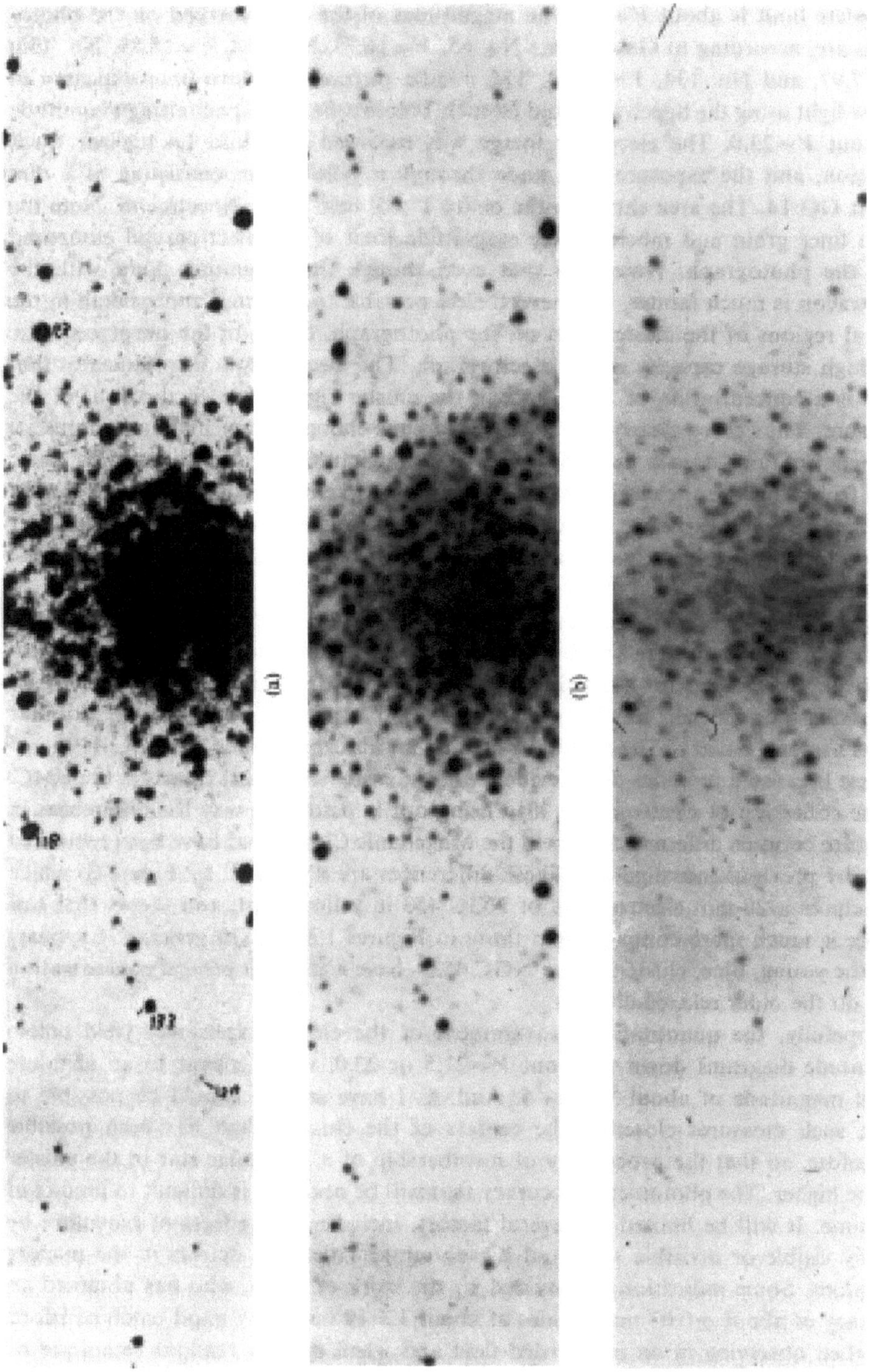

(a) (b) (c)

Fig. 1 a, b, c.

The plate limit is about $V=21$. The magnitudes of the stars marked on the photograph are, according to Gascoigne,: No. 65, $V=16.73$, No. 118, $V=16.89$, No. 133, $V=17.97$, and No. 134, $V=19.54$. The middle picture is a three-hour exposure in yellow light using the Spectracon and 60-inch Tololo reflector. The limiting magnitude is about $V=23.0$. The electronic image was recorded on Ilford L4 nuclear track emulsion, and the exposure was made through a yellow filter consisting of 2 mm Schott GG 14. The area shown is the entire $1' \times 5'$ field of the Spectracon. Note the much finer grain and much fainter magnitude limit of the electrograph compared with the photograph. Note also that even though the magnitude limit with the Spectracon is much fainter, it is nevertheless possible to see much more detail in the central regions of the cluster than on the photograph, owing to the linear response and high storage capacity of the electrograph. The electrograph also indicates that there is a concentration of faint stars in the cluster right down to the limit of the exposure. Thus there clearly is not, as Gascoigne thought from his photographs, a deficiency of faint stars in the cluster. The bottom picture in Figure 1 is a rather dense print from a one-hour electrograph in yellow light, taken through rather thick clouds. On this less dense exposure, the distribution of the brightest red giants in the cluster is seen to be much less concentrated to the center of the cluster than is that of the fainter stars.

Figure 2 shows a one-hour electrograph of NGC 419 in yellow light. The electrograph brings out a feature which has apparently gone unnoticed on previous photographs of the cluster: the obscuration of the righthand portion of the cluster by an interstellar absorption cloud. Gascoigne has found a similar case among the globular clusters in our Galaxy. The detection of such an absorbing cloud in NGC 419 is of interest because it provides direct evidence of the existence of dust clouds in the SMC.

The collection of electrographs also bring out in a striking way the differences in structure between different clusters in the Magellanic Clouds that have been remarked upon by previous investigators. These differences are illustrated by Figure 3, which reproduces a 20-min electrograph of NGC 453 in yellow light, and shows that this cluster is much more compact than those in Figures 1 and 2. In general, it appears that the young, blue, clusters – like NGC 453 – have a stronger central concentration than do the older relaxed clusters.

Hopefully, the quantitative measurement of the electrographs will yield color-magnitude diagrams down to about $V=22.5$ or 23.0, which means to an absolute visual magnitude of about $M_V=+4$. And, as I have said, it should be possible to make such measures closer to the centers of the clusters than has been possible heretofore, so that the probability of membership of a particular star in the cluster will be higher. The photometric accuracy that will be obtained is difficult to predict at this time. It will be limited by several factors, including the effects of crowding by nearby visible or invisible stars and by inhomogeneities and defects in the nuclear emulsions. Some indication is provided by the work of Kron, who has obtained an accuracy of about ± 0.03 magnitudes at about $V=19$ on a very good batch of Ilford G5 when observing in an uncrowded field and when using a readout technique of

Fig. 2.

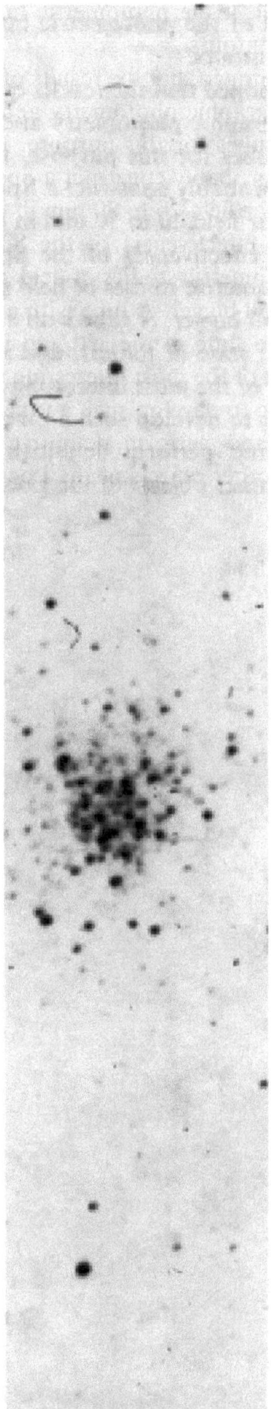

Fig. 3.

extracting all of the photometric information from the star image by measuring its isodensity contours.

It is to be hoped that the results of this program will help to demonstrate the merits of electronographic photometry and give some impetus towards the construction of larger-field tubes for this purpose. Experiments made by Professor McGee indicate that he can probably construct a Spectracon with a field of 15×30 mm, and perhaps even a circular field 20 to 30 mm in diameter. Such a field would already enormously enhance the effectiveness of the Spectracon for stellar photometry. However, for general photometric studies of field stars and extended objects one would like to have something still bigger. A tube with a cathode of about 10 cm diameter now appears to be within the state of the art, and in fact Professor Lallemand being, as always, in the forefront of the most interesting instrumental developments, is now in the process of attempting to develop such a tube. It is our hope to return in a few years time with this device and perform definitive photometric measurements in the Magellanic Clouds and other objects of the Local Group.

APPLICATION OF OBJECTIVE PRISM TECHNIQUES IN THE MAGELLANIC CLOUDS

JÜRGEN STOCK

Universidad de Chile, Observatorio Astronómico Nacional, Santiago de Chile

The Magellanic Clouds offer the possibility to investigate the stellar content and distribution within two entire stellar systems. For that purpose it is necessary to separate member stars from foreground objects, and subsequently determine the types of the member stars. Theories of stellar evolution as well as those concerning the formation of systems of stars and their kinematic and dynamic properties have been largely influenced by the characteristic distribution and frequency of different types of stars within our galaxy and other systems. Consequently, a classification of member stars of the Magellanic Clouds must be detailed enough to permit the segregation of groups of different evolutionary or kinematic significance. It shall be shown that existing objective prism techniques may well serve such purpose.

It has been shown by Slettebak and Stock that low-dispersion objective prism spectra taken with a UV-transparant prism are suitable for a two-dimensional classification in certain areas of the HR-diagram. This technique was broadened to other types by Upgren, and is now expanded again by the author.

Before entering into a description of the spectral features used for classification we must first discuss which limiting magnitude and thus which limiting distance modulus we expect to reach. The limitation for a given telescope and prism is determined by two factors: (1) The background brightness determines the limiting magnitude that can be reached. (2) Their angular size determines the number of overlapping spectra which are unsuitable for classification. Plates are on display which reach the limiting photographic magnitude $14\overset{m}{.}5$ in the area of the Large Magellanic Cloud. This means that on these plates we reach stars brighter than absolute magnitude -4.0 within the cloud. Several steps can be taken to reach fainter stars. As can be seen on the plates on display, particularly the luminosity criteria for certain stars could be recognized in spectra of considerably lower dispersion. Also, these criteria are restricted to a rather limited spectral range. Plates can be taken through a suitable filter, thus reducing background brightness and overlaps. One plate taken through a filter is also on display. It is estimated that the dispersion could be reduced by about a factor three, accompanied by a reduction in width of about a factor two. This corresponds to a gain of nearly two magnitudes. The gain obtained by using a filter is estimated to be about one magnitude. Thus, the limit of significant spectral classification seems to be between the 17th and 18th apparent magnitude. We should be able to reach about the A0-main sequence stars in the Large Cloud.

We must now consider whether the low dispersion proposed really permits a significant classification. The most suitable equipment available at present is the

Muller (ed.), The Magellanic Clouds, 181–183. *All Rights Reserved.*

24-inch Curtis Schmidt telescope on Cerro Tololo with a 4.5 degree prism on loan from the Warner and Swasey Observatory. The dispersion at Hγ is 625 Å/mm. An unreddened early-type spectrum of correct exposure extends to about 3300 Å with usable density. Early-type stars show spectra with no lines or very weak hydrogen lines. Some do present a continuous emission feature in the region of the Balmer continuum. Such stars are classified as OB or OBce depending on whether they show the emission. They comprise a large variety of stars with absolute magnitudes ranging from about -4 to -8, thus not forming a sufficiently well defined natural group, except that all stars included are in their early stages of evolution. From B 2 to B 4 the He- lines are visible. Supergiants can be separated on the basis of the weaker hydrogen lines. We expect that with the proposed lower dispersion all stars from O to B 4 will merge into one single OB-group of consequently even less definition. From B 5 on the Balmer discontinuity becomes visible. The amplitude of the Balmer discontinuity serves mainly as a temperature criterion, while its sharpness serves as a luminosity criterion. In underexposed spectra the intensity of the Balmer lines takes the role of the amplitude of the Balmer discontinuity criterion. However, in that case certain types of sub-luminous stars will merge with the group of supergiants. The criteria mentioned permit to classify in steps of two spectral subclasses, and furthermore permit to subdivide the range from luminosity class Ia^+ to V into three steps at B 5, and into eight steps at A 0. From A 0 on the Ca II-lines H and K serve as a temperature criterion, suitable up to late F-stars. For later A-stars and F-stars the luminosity criterion, that is, the sharpness of the Balmer discontinuity becomes insensitive at the lower luminosities, but continues to be suitable for the segregation of supergiants up to about G 5. With the proposed lower dispersion criteria based on spectral lines will become less sensitive, but those related to the appearance of the Balmer discontinuity may well be even more apparent. In well exposed or overexposed spectra there is an additional criterion in the far ultraviolet which seems to be very temperature sensitive. From G 0 on there appear a large number of new features in the UV, the significance of which is under study now. At present we can only say that these criteria cannot be fitted into a two-dimensional system, not even into a system of three dimensions. A large number of plates covering more than four thousand square degrees and reaching the 14th magnitude is available for further study. These plates cover areas from galactic latitude zero to the South Galactic Pole. Furthermore, several pairs of plates with inverted direction of the dispersion have been taken. These pairs of plates permit to segregate stars of large radial velocity. It is expected that from this material we can derive some conclusions concerning the significance of the different criteria visible in the ultraviolet of the spectra.

Now some words about the plates on exhibit. These plates were obtained just a few weeks ago. Consequently their analysis is far from terminated. On the cover plates all supergiants from B 5 to G 5 were marked. It is obvious that the surface distribution of these stars is different from that of the remainder, and actually the majority of the bright member stars. These belong to the class OB. The distribution of the b- to g-supergiants does not follow the distribution of the luminous matter either which consists of unresolved faint stars, nebulosities, etc., The faintest supergiants belong to the luminosity class Ib or II.

Apparently the b- to g-supergiants of the Large Cloud are more dispersed than the OB-stars. This is similar to what we are finding in the Milky Way. Deep plates taken along the fringes of the Milky Way show more faint supergiant stars of the types mentioned than OB-stars. This fact is not difficult to understand on the basis of the current theories of stellar formation and evolution.

SUMMARY AND DESIDERATA

J. H. OORT

University Observatory, Leiden, The Netherlands

I have given several summaries in my life, but for the present one I feel in as bad a position as Mrs. Burbidge, because this is also my first talk on the Magellanic Clouds, and because I am far from being an expert. I have made no observations, and have not even attempted to do anything about the dynamical theory of the Clouds. But the worst thing is that I have not really thought enough about the subject to give a meaningful summary.

The principal qualification that I could have for speaking on this subject is the comparison with our own Galaxy. But there a bad point comes in. I have been very much interested in the general features of the Galactic System but in particular also in its nuclear region. Now one of the important negative characteristics of the Magellanic Clouds is that they have no pronounced nuclear region. Dr. Mathewson has suggested this morning that perhaps the 30 Doradus complex would be similar to Sagittarius A, but this does not imply that it can be considered as the nucleus of the Large Cloud. It certainly does not *appear* to be so, as it is located rather excentrically. Moreover, the essential thing of the nucleus of our own Galaxy is not perhaps this Sagittarius A but the enormous concentration of mass density. This is not found around 30 Doradus.

One might say that in galaxies with well developed nuclear regions things have become more regular, as a consequence of this general concentration of mass towards the centre. In objects like the Magellanic Clouds, without such a central government, it seems that things go less well. The different provinces and states get too independent, and develop all on their own, which makes things complicated, and very difficult to summarize.

Many of the contributions that have been made to this Symposium have been in the nature of summaries of existing data. I certainly do not want to try to repeat these fine introductory reports, but want principally to comment on some general features.

It has been mentioned that the Large Cloud is possibly a very-late-type and quite irregular barred spiral. Superficially, the Small Cloud looks much like the Large Cloud, and it has been suggested that the elongated structure of the Small Cloud might also be something like a bar. I do not think the evidence for that is very convincing. Certainly, the population content of the bar in the Large Cloud appears to be quite different from that in the Small Cloud. In the Large Cloud neither neutral nor ionized hydrogen, nor early type stars are concentrated in the bar, while in the Small Cloud the elongated structure is just the place where the hydrogen and the early-type stars are mostly concentrated.

The total gas content of the two Clouds has been discussed many times during these two days. There seems to be fairly strong evidence that at least in the Small Cloud the

Muller (ed.), The Magellanic Clouds, 184–189.

gas constitutes of the order of $\frac{1}{3}$ of the total mass, and perhaps even more. In the Large Cloud a figure of 9% was shown in the table that Dr. Kerr put on the blackboard this morning, but, as has been remarked, this depends largely on the assumed inclination. I suppose this number was computed with the more or less conventional inclination of about 25° to the plane of the sky. Personally, I have the impression that there is no sufficient evidence for so small an inclination, and that it might as well be closer to a normal value, say of the order of 45°. In that case the total mass would go down by a factor of 3 or so, and the hydrogen content would go up to about the same fraction of the total mass as for the Small Cloud. Now, whether or not this is true, it is in any case so that the hydrogen content is large in both Clouds; considerably larger than in the Galactic System, where it is only about 3% of the mass. One might well ask why there is this large difference between the Clouds and the Galactic System. In a way one might say that this is mainly due to the large concentration of mass towards the central parts of the Galactic System, which, as I have mentioned, does not exist in the Magellanic Clouds. The latter may be more comparable with the outer parts of the Galaxy. In the surroundings of the sun about 20% of the total mass per unit of volume is gaseous, which is not too different from what we find in the Magellanic Clouds.

Another question that might be asked in connection with the gas is whether the ratio of dust to gas is smaller in the Magellanic Clouds than it is in the Galactic System in the general vicinity of the sun. From the evidence testifying on the smallness of the absorption within the Clouds it looks a bit that way, but I would not understand at all why this should be so.

The chemical composition has been discussed by Dr. Burbidge this afternoon. It appears to be difficult to say anything with great certainty about differences between the chemical composition of the Clouds and the Galactic System.

Let us now discuss for a minute those seemingly independent provinces within the Clouds. As we have heard from Dr. Kerr, the Large Cloud is extremely patchy; some 52 large H I concentrations have been noted in it, very large complexes with masses of more than a million solar masses each. Among them is one outstanding concentration, the 30 Doradus complex, with dimensions of 3000 by 1000 parsecs, and a hydrogen mass in the neighborhood of 10^8 solar masses. I do not think there is an essential difference in patchiness with the region of the Galactic System that surrounds us, where we also find large concentrations of hydrogen in the spiral arms; however, no complex of the size and mass of the 30 Doradus complex is known in our own Galaxy.

In general, the distribution of gas in the Large Cloud gives evidence of concentration to a thin layer, but in addition, there are indications of other layers, or at least of large patches of gas which deviate from the main layer. This evidence comes mainly from the 21 cm line profiles. One might perhaps compare this outlying gas to a kind of halo around the Magellanic Clouds, possibly a halo bearing some resemblance to the halo of the Galactic System as it might have been at the time of its formation.

There are pronounced differences between the two Clouds, as we have amply seen in the discussions. For instance, there is a difference in the patchiness. The Small Cloud shows much less of this patchy distribution than the Large Cloud. There is also a pronounced difference in the number of bright supergiants. There are about 10 times as many very bright stars in the Large Cloud than in the Small Cloud, notwithstanding the fact that the total amount of gas in the two Clouds is very nearly the same. So, in some still obscure way, as Mrs. Burbidge has already indicated, there must at present be special conditions in the Large Cloud, which favour the birth of very massive stars.

One of the most interesting problems, also for the future, would seem to be to make out in greater detail what the bar in the Large Cloud actually consists of. We have already seen that red giants seem to be concentrated in the bar (c.f. the beautiful pictures displayed by Bok). These stars have sometimes been referred to as population II objects, but I think that these bright supergiants should all be classed among population I. I shall return to the bar and to the birth of stars in a few minutes. But first I would like to call the attention to another phenomenon in the two Magellanic Clouds which is extremely interesting, and which we have not yet been able to study in the same way in the Galactic System, or in other systems, namely the phenomenon of the huge shell structures, the first of which was found by Mathewson and Westerlund in the Large Cloud. If it is a real shell, as it looks to be, it must be a thing of phantastic mass. Three gaseous shell-like structures were subsequently discovered in the Small Cloud, by Hindman. Here, I think, the evidence is rather convincing that we are actually dealing with giant shell structures of enormous masses, of the order of 10^7 solar masses, showing expansion velocities of the order of 20 km/sec and radii in the order of 700 parsec. These are phantastically interesting objects. It would be exciting to see if stars could be found associated with the shells, and whether thereby something could be learned about the star birth process within these shells.

Dr. Gascoigne has reported on the various types of globular and other clusters observed in the Magellanic Clouds. The most striking and unexpected thing here was the discovery, already quite a while ago, of the white globular clusters: clusters that are as massive as ordinary globular clusters in our own Galaxy, but which seem to be quite young. One might well ask what the reason can be that the formation of such globular clusters in the Magellanic Clouds still continues at the present time while it does not do so in our own Galaxy. This may well be due to a difference in the general state of motion of the gas. It may be that the differential rotation of the Galaxy precludes the formation of such huge clusters at the present time, while in the early history of our Galaxy the conditions in its halo were apparently favorable for the formation of big objects.

Some people have reasoned that the globular clusters may not have been formed within the galaxies but had existed already prior to the formation of galaxies, being later taken along with the mass concentrations in the galaxies. To me this does not seem probable, because, though there are in the galactic halo clusters of low metal content, there are also quite a number in which the metal content is quite high. It is likely that these metals would have been formed in the early stages of the history of our Galaxy by star formation and super-nova explosions in that formation period. So, personally, I would rather think of at least most of the globular clusters as having been formed within the galaxies than in a much earlier stage of the Universe.

Now I want to say a few words about the rotation and the age of the Clouds. I have the impression, from what Dr. Kerr said this morning that there is no relation between the direction of rotation of the two Clouds. This would not be surprising, because in most double galaxies the direction of the rotation is not correlated for the two components. Concerning the axis of rotation of the Large Cloud one may ask whether there is any evidence that it is perpendicular to the bar of the Large Cloud, as we believe to be the case for most barred spirals. I do not think there is evidence to contradict such an assumption.

As far as the time scale is concerned, and the apparent youth of the Magellanic Clouds, it is to be noted that the period of rotation is similar to the period of rotation of the Galactic System in our neighborhood. Why is it then that the Clouds are so much more poorly organized? If the periods of rotation had been much longer, or the ages much younger, one might think that the Clouds would not have had enough time to organize themselves, not having made enough revolutions. It is difficult to assume that the Magellanic Clouds would be very considerably younger than the Galactic System. After all we find RR Lyrae variables and fairly old globular clusters in them. One might possibly think of a factor of two or so in the age, but not much more. Even if you make them only 5 billion years old instead of 10 billion, there would still have been many revolutions and at first sight one might think that there would have been time for the gas to re-arrange itself in a more symmetrical pattern. It may again, I think, have been the lack of the existence of a proper nuclear region, and of a proper concentration of mass near it, which has prevented such a re-arrangement. In the main volume of the Clouds there seems to be very little differential rotation; in such a case, such unevenness as we see now could be conserved for very long periods. So I do not think there is much reason to worry about this, or to conclude that they are probably young systems.

Dr. Burbidge has raised the problem whether they have belonged to our Galaxy for a long time, or whether they are things that have fallen in recently, or possibly have been expelled from the nucleus or some other part of our Galaxy. I do not think that at present one can say anything interesting about this question. We can certainly not exclude the possibility that their closeness to our own Galaxy is only a very recent phenomenon, because even the Andromeda nebula and our Galaxy seem at present to be falling towards each other for the first time, and so there is no reason to assume that the Local Group has already gathered long ago. This might also apply to the Magellanic Clouds. But there is no positive evidence for it.

As regards the gravitational fields in the Clouds, if the gas content is really as high as $\frac{1}{3}$ or more of the total mass, and we see how extremely irregular, especially in the Large Cloud, the distribution of gas is, we must conclude that also the gravitational field will show considerable large-scale irregularities. Even the older type stars might not show too regular a structure in such an irregular and asymmetric gravitational field.

Now I would like to say a few words about the distribution of individual stars, on which we have had so many discussions. In the first place, of course, it is essential to be able to distinguish foreground stars and members of the Clouds. Dr. Fehrenbach and his coworkers have made a very important contribution with their objective prism velocities to make this distinction in a very definite way for the brighter members of the Clouds. One also needs, of course, to make this separation for the fainter stars. There it is more difficult, because there are more fainter foreground stars, and it is harder to get radial velocities. But fortunately, as Dr. Walraven has shown in his talk, it is possible to make this distinction almost completely by means of multi-colour, and even five-colour photometry. I suppose that in the future photometric work on the Clouds will be largely done by photometry with even more channels than in the photometer he has described, and this will certainly enable one completely to separate the actual members from foreground stars.

Making multi-colour observations of the fainter stars is evidently one of the big desiderata for future work on the Clouds. If I may make a guess, UBV photometry will soon become obsolete, and will entirely be replaced by multi-colour work, which you can get for the same money, and which will evidently yield much more information.

We have heard a great deal about the distribution of different types of stars. It is evident that the supergiant stars are directly related to the hydrogen distribution, and so are, of course, the emission nebulae that go with them. One particularly intriguing question in this connection, that has also been touched upon by Mrs. Burbidge, is how this star formation seems to go on locally for a certain period and then appears to stop. The most striking instance is the bar of the Large Cloud. As the Gaposchkins have pointed out there is an enormous concentration of cepheids in the bar. These may have ages of 10^8 yr, or thereabouts. This means that as short ago as 10^8 yr there must have been an enormous burst of star formation in this bar structure of the Large Cloud. One wonders where the hydrogen that must have given rise to the birth of these stars has gone in the meantime, because there is now no concentration of hydrogen at all in the bar. It is unlikely that it could have been used up in star formation, but the only other alternative would be that it has been blown away completely, and that is not a small matter, because the bar has a total length of about 3 kiloparsec.

I turn now to the investigations of the stars themselves. We have heard a number of most interesting reports on this subject. There has been, first, the report by Dr. Walraven on the information which the multi-colour observations yield about the structure of the atmospheres of the supergiants. There has been an extremely interesting talk by Dr. Christy this morning on the properties of the cepheids and the mechanism of cepheid pulsation. From this talk, as well as that by Dr. Walraven, it is abundantly clear how important the Magellanic Clouds are for this kind of semi- and purely-theoretical investigations; there is no place in the Galaxy where you can study these stars so completely as in the Magellanic Clouds. In the same line we have heard the enthusiastic account by Dr. Kippenhahn this afternoon. I have been very much impressed by the manner in which he, and Dr. Demarque and Dr. Burbidge, have presented these very difficult subjects; knowing very little about them, I have the impression that I now understand at least a little bit, and that, I think, is quite a compliment to the people who have given these introductions.

I have not yet mentioned many *desiderata.* The most important desideratum I can think of is to get good age determinations for stars in various parts of both Clouds, because nowhere can we see better the whole process of large-scale formation of stars, beginning with the assembling of interstellar gas into large complexes, and ending by its disappearance. At present this subject is still in its infancy, because most investigations have been concerned only with the brightest supergiants, the very youngest stars. It is evidently enormously interesting to follow this up to somewhat older stars, to see how the stars move away from their place of birth and to see also how the gas behaves after such bursts of star formation have taken place. This is still quite enigmatic, as I pointed out already in the case of the bar in the Large Cloud. But there is also the problem of the comparison between the Small Cloud and the Large Cloud. Why has the Large Cloud so many more very young stars than the Small Cloud? These are all questions we can hope to get much more insight in in a near future. The method described by professor Strömgren, with which he promised us that he could get reliable age determinations as soon as he gets a sufficiently large instrument at his disposal will certainly take an important place in these investigations.

To conclude I would stress that it is evidently of extreme importance for the work in the Clouds to go down to still fainter limits than we can reach at the present time. I would like to refer in this connection to the introduction given by Dr. Walker on image tubes, and on the plans of Dr. Lallemand to provide us with extremely beautiful new instruments for the future.

1 August 1969

ASTROPHYSICS AND SPACE SCIENCE LIBRARY

Edited by

J. E. Blamont, R. L. F. Boyd, L. Goldberg, C. de Jager, Z. Kopal, G. H. Ludwig, R. Lüst, B. M. McCormac, H. E. Newell, L. I. Sedov, Z. Švestka, and W. de Graaff

1. C. de Jager (ed.), *The Solar Spectrum. Proceedings of the Symposium held at the University of Utrecht, 26–31 August, 1963.* 1965, XIV + 417 pp.

2. J. Ortner and H. Maseland (eds.), *Introduction to Solar Terrestrial Relations. Proceedings of the Summer School in Space Physics held in Alpbach, Austria, July 15–August 10, 1963 and Organized by the European Preparatory Commission for Space Research.* 1965, IX + 506 pp.

3. C. C. Chang and S. S. Huang (eds.), *Proceedings of the Plasma Space Science Symposium, held at the Catholic University of America, Washington, D.C., June 11–14, 1963.* 1965, IX + 377 pp.

4. Zdeněk Kopal, *An Introduction to the Study of the Moon.* 1966, XII + 464 pp.

5. Billy M. McCormac (ed.), *Radiation Trapped in the Earth's Magnetic Field. Proceedings of the Advanced Study Institute, held at the Chr. Michelsen Institute, Bergen, Norway, August 16–September 3, 1965.* 1966, XII + 901 pp.

6. A. B. Underhill, *The Early Type Stars.* 1966, XIII + 282 pp.

7. Jean Kovalevsky, *Introduction to Celestial Mechanics.* 1967, VIII + 427 pp.

8. Zdeněk Kopal and Constantine L. Goudas (eds.), *Measure of the Moon. Proceedings of the Second International Conference on Selenodesy and Lunar Topography held in the University of Manchester, England, May 30–June 4, 1966.* 1967, XVIII + 479 pp.

9. J. G. Emming (ed.), *Electromagnetic Radiation in Space. Proceedings of the Third ESRO Summer School in Space Physics, held in Alpbach, Austria, from 19 July to 13 August, 1965.* 1968, VIII + 307 pp.

10. R. L. Carovillano, John F. McClay, and Henry R. Radoski (eds.), *Physics of the Magnetosphere. Based upon the Proceedings of the Conference held at Boston College, June 19–28, 1967.* 1968, X + 686 pp.

11. Syun-Ichi Akasofu, *Polar and Magnetospheric Substorms.* 1968, XVIII + 280 pp.

12. Peter M. Millman (ed.), *Meteorite Research. Proceedings of a Symposium on Meteorite Research held in Vienna, Austria, 7–13 August, 1968.* 1969, XV + 941 pp.

13. Margherita Hack (ed.), *Mass Loss from Stars. Proceedings of the Second Trieste Colloquium on Astrophysics, 12–17 September, 1968.* 1969, XII + 345 pp.

14. N. D'Angelo (ed.), *Low-Frequency Waves and Irregularities in the Ionosphere. Proceedings of the 2nd ESRIN-ESLAB Symposium, held in Frascati, Italy, 23–27 September, 1968.* 1969, VII + 218 pp.

15. G. A. Partel (ed.), *Space Engineering. Proceedings of the Second International Conference on Space Engineering, held at the Fondazione Giorgio Cini, Isola di San Giorgio, Venice, Italy, May 7–10, 1969.* 1970, XI + 728 pp.

16. S. Fred Singer (ed.), *Manned Laboratories in Space. Second International Orbital Laboratory Symposium.* 1969, XIII + 133 pp.

17. B. M. McCormac (ed.), *Particles and Fields in the Magnetosphere. Symposium Organized by the Summer Advanced Study Institute, held at the University of California, Santa Barbara, Calif., August 4–15, 1969.* 1970, XI + 450 pp.

18. Jean-Claude Pecker, *Experimental Astronomy.* 1970, X + 105 pp.

19. V. Manno and D. E. Page (eds.), *Intercorrelated Satellite Observations related to Solar Events. Proceedings of the Third ESLAB/ESRIN Symposium held in Noordwijk, The Netherlands, September 16–19, 1969.* 1970, XVI + 627 pp.

20. L. Mansinha, D. E. Smylie and A. E. Beck, *Earthquake Displacement Fields and the Rotation of the Earth. A NATO Advanced Study Institute. Conference Organized by the Department of University of Western Ontario, London, Canada, 22 June–28 June, 1969.* 1970, XI + 308 pp.

21. Jean-Claude Pecker, *Space Observatories.* 1970, XI + 120 pp.

22. L. N. Mavridis (ed.), *Structure and Evolution of the Galaxy. Proceedings of the NATO Advanced Study Institute, held in Athens, September 8–19, 1969.* 1971, VII + 312 pp.

24. B. M. McCormac (ed.), *The Radiating Atmosphere. Proceedings of a Symposium Organized by the Summer Advanced Study Institute, held at Queen's University, Kingston, Ontario, August 3–14, 1970.* 1971, XI + 455 pp.

In preparation:

25. G. Fiocco (ed.), *Mesospheric Models and Related Experiments. Proceedings of the 4th ESRIN-ESLAB Symposium, held at Frascati, Italy, July 6–10, 1970.*

26. I. Atanasijević, *Selected Exercises in Galactic Astronomy.*

27. C. J. Macris (ed.), *Physics of the Solar Corona. Proceedings of NATO Advanced Study Institute on Physics of the Solar Corona, held at Cavouri-Vouliagmeni, Athens, Greece, 6–17 September 1970.*

SOLE DISTRIBUTORS FOR U.S.A. AND CANADA:

SPRINGER-VERLAG NEW YORK, INC., 175 Fifth Ave., New York, N.Y. 10011

MIX
Papier aus verantwortungsvollen Quellen
Paper from responsible sources
FSC® C105338

If you have any concerns about our products,
you can contact us on
ProductSafety@springernature.com

In case Publisher is established outside the EU,
the EU authorized representative is:
Springer Nature Customer Service Center GmbH
Europaplatz 3, 69115 Heidelberg, Germany

Printed by Libri Plureos GmbH
in Hamburg, Germany